C000040643

LECTURES ON ORDINARY DIFFERENTIAL EQUATIONS

LECTURES ON ORDINARY DIFFERENTIAL EQUATIONS

WITOLD HUREWICZ

DOVER PUBLICATIONS, INC.
MINEOLA, NEW YORK

Bibliographical Note

This Dover edition, first published in 1990 and reissued in 2014, is an unaltered republication of the first M.I.T. Press paperback edition (1964) of the work first published by the M.I.T. Press, Cambridge, Mass., in 1958.

Library of Congress Cataloging-in-Publication Data

Hurewicz, Witold, 1904–1956.
    Lectures on ordinary differential equations / Witold Hurewicz. — Dover ed.
        p. cm.
    Reprint. Previously published: Cambridge, Mass. : M.I.T. Press, 1964.
    Includes bibliographical references and index.
    ISBN-13: 978-0-486-66420-0 (pbk.)
    ISBN-10: 0-486-66420-1 (pbk.)
    1. Differential equations. I. Title.

QA372.H93 1990
515'.35—dc20

90-37994

Manufactured in the United States of America
66420102   2022
www.doverpublications.com

Preface

This book is a reprinting, with minor revisions and one correction, of notes originally prepared by John P. Brown from the lectures given in 1943 by the late Professor Witold Hurewicz at Brown University. They were first published in mimeographed form by Brown University in 1943, and were reissued by the Mathematics Department of The Massachusetts Institute of Technology in 1956. They are now reprinted with the permission of Brown University and of Dr. Stefan Hurewicz.

An appreciation of Witold Hurewicz by Professor Solomon Lefschetz, which first appeared in the *Bulletin of the American Mathematical Society*, is included in this book, together with a bibliography of his published works.

Since this book treats mainly of existence theorems, linear systems, and geometric aspects of nonlinear systems in the plane, a selected list of books on differential equations has been placed at the end of the volume for those interested in further reading.

The work of the several mathematicians who prepared this new edition was supported by the Office of Naval Research.

NORMAN LEVINSON

Massachusetts Institute of Technology

Witold Hurewicz
In Memoriam[*]

SOLOMON LEFSCHETZ

Last September sixth was a black day for mathematics. For on that day there disappeared, as a consequence of an accidental fall from a pyramid in Uxmal, Yucatan, Witold Hurewicz, one of the most capable and lovable mathematicians to be found anywhere. He had just attended the International Symposium on Algebraic Topology which took place during August at the National University of Mexico and had been the starting lecturer and one of the most active participants. He had come to Mexico several weeks before the meeting and had at once fallen in love with the country and its people. As a consequence he established from the very first a warm relationship between himself and the Mexican mathematicians. His death caused among all of us there a profound feeling of loss, as if a close relative had gone, and for days one could speak of nothing else.

Witold Hurewicz was born on June 29, 1904, in Lodz, Russian Poland, received his early education there, and his doctorate in Vienna in 1926. He was a Rockefeller Fellow in 1927–1928 in Amsterdam, privaat docent there till 1936 when he came to this country. The Institute for Advanced Study, the University of North Carolina, Radiation Laboratory and Massachusetts Institute of Technology (since 1945) followed in succession.

[*] Reprinted by permission from the *Bulletin of the American Mathematical Society*, vol. 63, no. 2, pp. 77–82 (March, 1957).

Mathematically Hurewicz will best be remembered for his important contributions to dimension, and above all as the founder of homotopy group theory. Suffice it to say that the investigation of these groups dominates present day topology.

Still very young, Hurewicz attacked dimension theory, on which he wrote together with Henry Wallman the book *Dimension theory* [39].[1] We come to this book later. The Menger-Urysohn theory, still of recent creation, was then in full bloom, and Menger was preparing his book on the subject. One of the principal contributions of Hurewicz was the extension of the proofs of the main theorems to separable metric spaces [2 to 10] which required a different technique from the basically Euclidean one of Menger and Urysohn. Some other noteworthy results obtained by him on dimension are:

(a) *A separable metric n-space (=n dimensional space) may be topologically imbedded in a compact metric n-space* [7].

(b) *Every compact metric n-space Y is the map of a compact metric zero-space X in such a manner that no point of Y has more than $n + 1$ antecedents, where n cannot be lowered, and conversely where this holds* dim $Y = n$. *In particular one may choose for X a linear set containing no interval* [6].

(c) Perhaps his best dimension result is his proof and extension of the imbedding theorem of compact spaces of dimension $\leq n$ in Euclidean E_{2n+1} which reads: *A compact metric n-space X may be mapped into E_{n+m} $(m = 1, 2, \cdots)$, so that the points which are images of k points of X make up a set of dimension $\leq n \div (k - 1)m$* [26].

This proposition may also be generalized as follows: *Any mapping f: $X \to E_{n+m}$ may be arbitrarily approximated by one behaving as stated.* Special case: *X may be mapped topologically into E_{2n+1}.* Earlier proofs of this last theorem existed. The wholly original proof of the main theorem by Hurewicz rests upon the utilization of the space E_{n+m}^{X} of mappings of $X \to E_{n+m}$, as defined by Fréchet and the proof that the mappings of the desired type are dense in E_{n+m}^{X}.

A more special but interesting dimensional result is:

(d) *Hilbert space is not a countable union of finite dimensional spaces* [10].

Recall R. L. Moore's noteworthy proposition: a decomposition of the two-sphere S^2 in upper semi-continuous continua which do not disconnect S^2 is topologically an S^2. Hurewicz showed [17] that *for S^3 no such result holds and one may thus obtain topologically any compact metric space.* This shows that R. L. Moore's results describe a very special property of S^2.

---

[1] Square brackets refer to the bibliography at the end.

Another investigation of Hurewicz marked his entrance into algebraic topology. The undersigned had introduced so called LC^n spaces: compact metric spaces locally connected in terms of images of p-spheres for every $p \leqq n$. One may introduce HLC^n spaces with images of p-spheres replaced by integral p-cycles and contractibility to a point by ~ 0 in the sense of Vietoris. Hurewicz proved this very unexpected property: *N.a.s.c. for X as above to be LC^n is HLC^n plus local contractibility of closed paths* [33]. An analogous condition will appear in connection with homotopy groups.

We come now to the four celebrated 1935 Notes on the homotopy groups, of the Amsterdam Proceedings [29; 30; 34; 35]. The attack is by means of the function spaces X^Y. Let Y be a separable metric space which is connected and locally contractible in the sense of Borsuk. Let S^p denote the p-sphere. Let x_0 be a fixed point of S^{n-1}, $n \geqq 1$, and y_0 a fixed point of Y. Let N be the subset of $Y^{S^{n-1}}$ consisting of the mappings F such that $Fx_0 = y_0$. The group of the paths of N is the same for all components of M. It is by definition the nth homotopy group $\pi_n(Y)$ of Y. For $n = 1$ it is the group of the paths of Y, and hence generally noncommutative but for $n > 1$ the groups are always commutative. Hurewicz proved the following two noteworthy propositions:

I. *When the first $n - (n \geqq 2)$ homotopy groups of the space Y (same as before) are zero then the nth $\pi_n(Y)$ is isomorphic with $H_n(Y)$, the nth integral homology group of Y.*

II. *N.a.s.c. for a finite connected polyhedron Π to be contractible to a point is $\pi_1(\Pi) = 1$ and $H_n(\Pi) = 0$ for every $n > 1$.*

For many years only a few homotopy groups were computed successfully. In the last five years however great progress has been made and homotopy groups have at last become computable mainly through the efforts of J.-P. Serre, Eilenberg and MacLane, Henri Cartan, and John Moore.

Many other noteworthy results are found in the four Amsterdam Proceedings Notes but we cannot go into them here. We may mention however the fundamental concept of homotopy type introduced by Hurewicz in the last note: Two spaces X, Y are said to be of the same homotopy type whenever there exist mappings $f: X \to Y$ and $g: Y \to X$ such that gf and fg are deformations in X and Y. This concept gives rise to an equivalence and hence to equivalence classes. This is the best known approximation to homeomorphism, and comparison according to homotopy type is now standard in topology. Identity of homotopy type implies the isomorphism of the homology and homotopy groups.

At a later date (1941) and in a very short abstract of this Bulletin [40] Hurewicz introduced the concept of exact sequence whose mushroom like expansion in recent topology is well known. The idea rests upon a collection of groups G_n and homeomorphism ϕ_n such that

$$\cdots \to G_{n+2} \xrightarrow{\phi_{n+1}} G_{n+1} \xrightarrow{\phi_n} G_n \to \cdots$$

and so that the kernel of ϕ_n is $\phi_{n+1}G_{n+2}$. This was applied by Hurewicz to homology groups and he drew important consequences from the scheme.

Still another noteworthy concept dealt with by Hurewicz is that of fibre space. In a Note [38] written in collaboration with Steenrod there was introduced the concept of the covering homotopy, its existence was established in fibre spaces, the power of the method was made clear. He returned to it very recently [45] to build fibre spaces on a very different basis. In another recent Note [46] written in collaboration with Fadell there was established the first fundamental advance beyond the theorem of Leray (1948) about the structure of spectral sequences of fibre spaces.

Hurewicz made a number of excursions into analysis, principally real variables. A contribution of a different nature was his extension of G. D. Birkhoff's ergodic theorem to spaces without invariant measure [42].

During World War II Hurewicz gave evidence of surprising versatility in distinguished work which he did for the Radiation Laboratory. This led among other things to his writing a chapter in the Servo Mechanisms series issued by the Massachusetts Institute of Technology.

The scientific activity of Hurewicz extended far beyond his written papers important as these may be. One way that it manifested itself is through his direct contact with all younger men about him. He was ready at all times to listen carefully to one's tale and to make all manner of suggestions, and freely discussed his and anybody else's latest ideas. One of his major sources of influence was exerted through his books. *Dimension theory* [39] already mentioned is certainly the definitive work on the subject. One does not readily understand how so much first rate information could find place in so few pages. We must also mention his excellent lectures on differential equations [41] which has appeared in mimeographed form and has attracted highly favorable attention.

On the human side Witold Hurewicz was an equally exceptional personality. A man of the widest culture, a first rate and careful linguist, one could truly apply to him *nihil homini a me alienum puto*.

Tales were also told of his forgetfulness—which made him all the more charming. Altogether we shall not soon see his equal.

BIBLIOGRAPHY

1. *Über eine Verallgemeinerung des Borelschen Theorems*, Math. Zeit. vol. 24 (1925) pp. 401–421.
2. *Über schnitte von Punktmengen*, Proc. Akad. van Wetenschappen vol. 29 (1926) pp. 163–165.
3. *Stetige bilder von Punktmengen.* I, Ibid. (1926) pp. 1014–1017.
4. *Grundiss der Mengerschen Dimensionstheorie*, Math. Ann. vol. 98 (1927) pp. 64–88.
5. *Normalbereiche und Dimensionstheorie*, Math. Ann. vol. 96 (1927) pp. 736–764.
6. *Stetige bilder von Punktmengen.* II, Proc. Akad. van Wetenschappen vol. 30 (1927) pp. 159–165.
7. *Verhalten separabler Räume zu kompakten Räumen*, Ibid. (1927) pp. 425–430.
8. *Über Folgen stetiger Funktionen*, Fund. Math. vol. 9 (1927) pp. 193–204.
9. *Relativ perfekte Teile von Punktmengen und Mengen*, Fund. Math. vol. 12 (1928) pp. 78–109.
10. *Über unendlich—dimensionale Punktmengen*, Proc. Akad. van Wetenschappen vol. 31 (1928) pp. 916–922.
11. *Dimension und Zusammenhangsstufe* (with K. Menger), Math. Ann. vol. 100 (1928) pp. 618–633.
12. *Über ein topologisches Theorem*, Math. Ann. vol. 101 (1929) pp. 210–218.
13. *Über der sogenannter Produktsatz der Dimensionstheorie*, Math. Ann. vol. 102 (1929) pp. 305–312.
14. *Zu einer Arbeit von O. Schreier*, Abh. Math. Sem. Hansischen Univ. vol. 8 (1930) pp. 307–314.
15. *Ein Theorem der Dimensionstheorie*, Ann. of Math. vol. 31 (1930) pp. 176–180.
16. *Einbettung separabler Räume in gleich dimensional kompakte Räume*, Monatshefte für Mathematik vol. 37 (1930) pp. 199–208.
17. *Über oberhalb-stetige Zerlegungen von Punktmengen in Kontinua*, Fund. Math. vol. 15 (1930) pp. 57–60.
18. *Theorie der Analytischen mengen*, Fund. Math. vol. 15 (1930) pp. 4–17.
19. *Dimensionstheorie und Cartesische Räume*, Proc. Akad. van Wetenschappen vol. 34 (1931) pp. 399–400.
20. *Une remarque sur l'hypothèse du continu*, Fund. Math. vol. 19 (1932) pp. 8–9.
21. *Über die henkelfreie Kontinua*, Proc. Akad. van Wetenschappen vol. 35 (1932) pp. 1077–1078.
22. *Stetige abbildungen topologischer Räume*, Proc. International Congress Zurich vol. 2 (1932) p. 203.
23. *Über Dimensionserhörende stetige Abbildungen*, J. Reine Angew. Math. vol. 169 (1933) pp. 71–78.
24. *Über Schnitte in topologischen Räumen*, Fund. Math. vol. 20 (1933) pp. 151–162.
25. *Ein Einbettungssatz über henkelfreie Kontinua* (with B. Knaster), Proc. Akad. van Wetenschappen vol. 36 (1933) pp. 557–560.

26. *Über Abbildungen von endlichdimensionalen Räumen auf Teilmengen Cartesischerräume*, Preuss. Akad. Wiss. Sitzungsber. (1933) pp. 754–768.

27. *Über einbettung topologischer Räume in cantorsche Mannigfaltigkeiten*, Prace Matematyczno-Fizyczne vol. 40 (1933) pp. 157–161.

28. *Satz über stetige Abbildungen*, Fund. Math. vol. 23, pp. 54–62.

29. *Höher-dimensionale Homotopiegruppen*, Proc. Akad. van Wetenschappen vol. 38 (1935) pp. 112–119.

30. *Homotopie und Homologiegruppen*, Proc. Akad. van Wetenschappen vol. 38 (1935) pp. 521–528.

31. *Über Abbildungen topologischer Räume auf die n-dimensionale Sphäre*, Fund. Math. vol. 24 (1935) pp. 144–150.

32. *Sur la dimension des produits cartésiens*, Ann. of Math. vol. 36 (1935) pp. 194–197.

33. *Homotopie, Homologie und lokaler Zusammenhang*, Fund. Math. vol. 25 (1935) pp. 467–485.

34. *Klassen und Homologietypen von Abbildungen*, Proc. Akad. van Wetenschappen vol. 39 (1936) pp. 117–126.

35. *Asphärische Räume*, Ibid. (1936) pp. 215–224.

36. *Dehnungen, Verkürzungen, Isometrien* (with H. Freudenthal), Fund. Math. vol. 26 (1936) pp. 120–122.

37. *Ein Einfacher Beweis des Hauptsatzes über cantorsche Mannigfaltigkeiten*, Prace Matematyczno-Fizyczne vol. 44 (1937) pp. 289–292.

38. *Homotopy relations in fibre spaces* (with N. E. Steenrod), Proc. Nat. Acad. Sci. U.S.A. vol. 27 (1941) pp. 60–64.

39. *Dimension theory* (with H. Wallman), Princeton University Press (Princeton Mathematical Series No. 4), 1941, 165 p.

40. *On duality theorems*, Bull. Amer. Math. Soc. Abstract 47-7-329.

41. *Ordinary differential equations in the real domain with emphasis on geometric methods*, 129 mimeographed leaves, Brown University Lectures, 1943.

42. *Ergodic theorem without invariant measure*, Ann. of Math. vol. 45 (1944) pp. 192–206.

43. *Continuous connectivity groups in terms of limit groups* (with J. Dugundji and C. H. Dowker), Ann. of Math. (2) vol. 49 (1948) pp. 391–406.

44. *Homotopy and homology*, Proceedings of the International Congress of Mathematicians, Cambridge, 1950, vol. 2, American Mathematical Society, 1952, pp. 344–349.

45. *On the concept of fiber space*, Proc. Nat. Acad. Sci. U.S.A. vol. 41 (1955) pp. 956–961.

46. *On the spectral sequence of a fiber space* (with E. Fadell), Proc. Nat. Acad. Sci. U.S.A. vol. 41 (1955) pp. 961–964; vol. 43 (1957) pp. 241–245.

47. Contributed Chapter 5, *Filters and servosystems with pulsed data*, pp. 231–261, in James, Nichols and Phillips *Theory of servomechanisms*, Massachusetts Institute of Technology, Radiation Laboratory Series, vol. 25, New York, McGraw-Hill, 1947.

48. *Stability of mechanical systems* (co-author H. Greenberg), N. D. R. C. Report, 1944 (to appear in the Quarterly of Applied Mathematics).

49. Four reports on servomechanisms for the Massachusetts Institute of Technology Radiation Laboratory.

50. *Dimension of metric spaces* (with C. H. Dowker), Fund. Math. vol. 43 (January, 1956) pp. 83–88.

Contents

I

Differential Equations
of the First Order
in One Unknown Function

PART A.
THE CAUCHY-EULER APPROXIMATION METHOD

1. Definitions. Direction Fields

By a *domain D* in the plane we understand a connected open set of points; by a *closed domain* or *region* $\bar{D}$, such a set plus its boundary points. The most general differential equation of the first order in one unknown function is

$$F(x, y, y') = 0 \tag{1}$$

where F is a single-valued function in some domain of its arguments. A differentiable function $y(x)$ is a solution if for some interval of x, $(x, y(x), y'(x))$ is in the domain of definition of F and if further $F(x, y(x), y'(x)) = 0$. We shall in general assume that (1) may be written in the *normal form*:

$$y' = f(x, y) \tag{2}$$

where $f(x, y)$ *is a continuous function of both its arguments simultaneously* in some domain D of the x-y plane. It is known that this reduction may

1

be carried out under certain general conditions. However, if the reduction is impossible, e.g., if for some (x_0, y_0, y_0') for which $F(x_0, y_0, y_0') = 0$ it is also the case that $[\partial F(x_0, y_0, y_0')]/\partial y' = 0$, we must for the present omit (1) from consideration.

A solution or integral of (2) over the interval $x_0 \leq x \leq x_1$ is a single-valued function $y(x)$ with a continuous first derivative $y'(x)$ defined on $[x_0, x_1]$ such that for $x_0 \leq x \leq x_1$:

(a) $(x, y(x))$ is in D, whence $f(x, y(x))$ is defined

(b) $y'(x) = f(x, y(x))$. (3)

Geometrically, we may take (2) as defining a *continuous direction field* over D; i.e., at each point P: (x, y) of D there is defined a line whose slope is $f(x, y)$; and an integral of (2) is a curve in D, one-valued in x and with a continuously turning tangent, whose tangent at P coincides with the direction at P. Equation (2) does not, however, define the most general direction field possible; for, if R is a bounded region in D, $f(x, y)$ is continuous in R and hence bounded

$$|f(x, y)| \leq M, \qquad (x, y) \text{ in } R \qquad (4)$$

where M is a positive constant. If α is the angle between the direction defined by (2) and the x-axis, (4) means that α is restricted to such values that

$$|\tan \alpha| \leq M$$

The direction may approach the vertical as P: (x, y) approaches the boundary C of D, but it cannot be vertical for any point P of D.

This somewhat arbitrary restriction may be removed by considering the *system* of differential equations

$$\frac{dx}{dt} = P(x, y), \qquad \frac{dy}{dt} = Q(x, y) \qquad (5)$$

where P and Q are the direction cosines of the direction at (x, y) to the x- and y-axes respectively and hence are continuous and bounded; and a solution of (5) is of the form

$$x = x(t), \qquad y = y(t), \qquad t_0 \leq t \leq t_1$$

which are the parametric equations of a curve L in D. We shall be able to solve (5) by methods similar to those which we shall develop for (2); hence we shall for the present consider only the theory of the equation (2).

2. Approximate Solutions of the Differential Equation

Since the physicist deals only with approximate quantities, an approximate solution of a differential equation is just as good for his purposes as an exact solution, provided the approximation is sufficiently close. The present section will formulate the idea of an approximate solution and prove that the approximation may be made arbitrarily close. The methods used will be practical even though rather crude. More important, the demonstration of the existence of approximate solutions will later lead to the proof of the existence and uniqueness of exact solutions.

Definition 1. Let $f(x, y)$ be continuous, (x, y) in some domain D. Let $x_1 \leq x \leq x_2$ be an interval. Then a function $y(x)$ defined on $[x_1, x_2]$ is a solution of $y' = f(x, y)$ up to the error ϵ if:

(a) *$y(x)$ is admissible; i.e., $(x, y(x))$ is in D, $x_1 \leq x \leq x_2$.*
(b) *$y(x)$ is continuous; $x_1 \leq x \leq x_2$.*
(c) *$y(x)$ has a piecewise continuous derivative on $[x_1, x_2]$ which may fail to be defined only for a finite number of points, say $\xi_1, \xi_2, \cdots, \xi_n$.*
(d) *$|y'(x) - f(x, y(x))| \leq \epsilon; \ x_1 \leq x \leq x_2, x \neq \xi_i, i = 1, \cdots, n.*

Theorem 1. Let (x_0, y_0) be a point of D, and let the points of a rectangle $R: |x - x_0| \leq a, |y - y_0| \leq b$ lie in D. Let $|f(x, y)| \leq M, (x, y)$ in R. Then if $h = min (a, b/M)$, there can be constructed an approximate solution $y(x)$ of

$$y' = f(x, y) \tag{1}$$

over the interval $|x - x_0| \leq h$, such that $y(x_0) = y_0$, where the error ϵ may be an arbitrarily small positive number. Observe that h is independent of ϵ.

Proof. The rectangle

$$S: |x - x_0| \leq h, \qquad |y - y_0| \leq Mh \tag{2}$$

is contained in R by the definition of h. See Fig. 1.
Let the ϵ of the theorem be given. Since $f(x, y)$ is continuous in S, it is

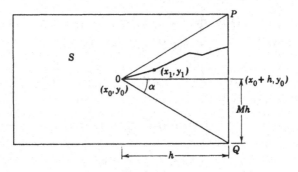

Fig. 1

uniformly so; i.e., given $\epsilon > 0$ (which we take to be the ϵ of the theorem) there exists $\delta > 0$ such that

$$|f(\tilde{x}, \tilde{y}) - f(x, y)| \leq \epsilon \qquad (3)$$

for $(\tilde{x}, \tilde{y})$; (x, y) in S; $|\tilde{x} - x|$, $|\tilde{y} - y| \leq \delta$.

Let $x_1, \cdots, x_{n-1}$ be any set of points such that:

(a) $x_0 < x_1 < x_2 < \cdots < x_{n-1} < x_n = x_0 + h$;

(b) $x_i - x_{i-1} \leq \min(\delta, \delta/M)$, $\qquad i = 1, \cdots, n$. $\qquad (4)$

We shall construct the approximate solution on the interval $x_0 \leq x \leq x_0 + h$; a similar process will define it on the interval $x_0 - h \leq x \leq x_0$.

The approximate solution will be a polygon constructed in the following fashion: from (x_0, y_0) we draw a segment to the right with slope of $f(x_0, y_0)$; this will intersect the line $x = x_1$ in a point (x_1, y_1). From (x_1, y_1) we draw a segment to the right with slope $f(x_1, y_1)$ intersecting $x = x_2$ at y_2; etc. The point (x_1, y_1) must lie in the triangle OPQ; for $\tan \alpha = M$, and $|f(x_0, y_0)| \leq M$. Likewise (x_2, y_2) lies in OPQ; etc. Hence the process may be continued up to $x_n = x_0 + h$, since the only way in which the process could stop would be for $f(x_k, y_k)$ to be undefined; in which case we should have $|y_k - y_0| > Mh$ contrary to construction. Analytically we may define $y(x)$ by the recursion formulas

$$y(x) = y_{i-1} + (x - x_{i-1})f(x_{i-1}, y(x_{i-1})) \qquad (5)$$

where

$$y_{i-1} = y(x_{i-1}), \qquad x_{i-1} \leq x \leq x_i, \qquad i = 1, \cdots, n$$

Obviously by definition $y(x)$ is admissible, continuous, and has a piecewise continuous derivative

$$y'(x) = f(x_{i-1}, y(x_{i-1})), \qquad x_{i-1} < x < x_i, \qquad i = 1, \cdots, n$$

which fails to be defined only at the points x_i, $i = 1, \cdots, n - 1$. Furthermore, if $x_{i-1} < x < x_i$

$$|y'(x) - f(x, y(x))| = |f(x_{i-1}, y_{i-1}) - f(x, y(x))| \qquad (6)$$

But by (4), $|x - x_{i-1}| < \min(\delta, \delta/M)$, and by (5)

$$|y - y_{i-1}| \leq M|x - x_{i-1}| \leq M\frac{\delta}{M} = \delta$$

Hence by (3)

$$|f(x_{i-1}, y_{i-1}) - f(x, y(x))| \leq \epsilon$$

and

$$|y'(x) - f(x, y(x))| \leq \epsilon \qquad x \neq x_i, \qquad i = 1, 2, \cdots, n - 1 \qquad (7)$$

Hence $y(x)$ satisfies all the conditions of Def. 1, and the construction required by the theorem has been performed. This method of constructing an approximate solution is known as the Cauchy-Euler method.

It is unnecessary to improve the value of h, since, in general, as we shall show, $y(x)$ is defined in a larger interval than $|x - x_0| \leq h$.

3. The Fundamental Inequality

With a certain additional restriction upon $f(x, y)$ we shall prove an inequality which will be the basis of our fundamental results.

Definition 2. *A function $f(x, y)$ defined on a (open or closed) domain D is said to satisfy Lipschitz conditions with respect to y for the constant $k > 0$ if for every x, y_1, y_2 such that $(x, y_1), (x, y_2)$ are in D*

$$|f(x, y_1) - f(x, y_2)| \leq k|y_1 - y_2| \tag{1}$$

In connection with this definition we shall need two theorems of analysis:

Lemma 1. *If $f(x, y)$ has a partial derivative for y, bounded for all (x, y) in D, and D is convex (i.e., the segment joining any two points of D lies entirely in D), then $f(x, y)$ satisfies a Lipschitz condition for y where the constant k is given by*

$$k = \text{l.u.b.} \left| \frac{\partial f(x, y)}{\partial y} \right|, \qquad (x, y) \text{ in } D \tag{2}$$

Proof. By Rolle's theorem there exists a number ξ such that

$$f(x, y_2) - f(x, y_1) = (y_2 - y_1) \frac{\partial f(x, \xi)}{\partial y}, \qquad y_1 \leq \xi \leq y_2$$

i.e.,

$$|f(x, y_2) - f(x, y_1)| \leq \text{l.u.b.} \left| \frac{\partial f(x, y)}{\partial y} \right| |y_2 - y_1|$$

since (x, ξ) is in D; whence the theorem.

Lemma 2. *If D is not convex, let D be imbedded in a larger domain D'. Let δ be the distance between the boundaries C and C' of D and D' respectively (i.e., $\delta = \min(\overline{PP'})$, P in C, P' in C'), and let $\delta > 0$. Then if $f(x, y)$ is continuous in D' (hence bounded by M, say, in D) and $\partial f/\partial y$ exists and is bounded by N in D', then $f(x, y)$ satisfies a Lipschitz condition in D with respect to y for a constant*

$$k = \max\left(N, \frac{2M}{\delta}\right)$$

Proof. Let P_1: (x, y_1) and P_2: (x, y_2) be in D.

(a) If $|y_1 - y_2| > \delta,$ $\left| \dfrac{f(x, y_1) - f(x, y_2)}{y_1 - y_2} \right| \le \dfrac{2M}{\delta}.$

(b) If $|y_1 - y_2| \le \delta,$ the segment $\overline{P_1 P_2}$ lies wholly in D'.

Hence as in Lemma 1, $\left| \dfrac{f(x, y_1) - f(x, y_2)}{y_1 - y_2} \right| \le N.$ Hence the lemma.

Theorem 2. *Let (x_0, y_0) be a point of a region R in which $f(x, y)$ is continuous and satisfies the Lipschitz condition for k. Let $y(x)$, $\tilde{y}(x)$ be admissible functions for $|x - x_0| \le h$ (where h is any constant, not necessarily that of Theorem 1), satisfying*

$$y' = f(x, y), \qquad |x - x_0| \le h$$

with errors ϵ_1 and ϵ_2 respectively. Set

$$p(x) = \tilde{y}(x) - y(x), \qquad \epsilon = \epsilon_1 + \epsilon_2 \tag{3}$$

Then

$$|p(x)| \le e^{k|x - x_0|} |p(x_0)| + \frac{\epsilon}{k} (e^{k|x - x_0|} - 1) \tag{4}$$

This is the fundamental inequality.

Proof. We give the proof only for $x_0 \le x \le x_0 + h$; a similar process will give the proof for $x_0 - h \le x \le x_0$. By Def. 1, except for a finite number of points

$$\left| \frac{dy}{dx} - f(x, y) \right| \le \epsilon_1, \qquad \left| \frac{d\tilde{y}}{dx} - f(x, \tilde{y}) \right| \le \epsilon_2, \qquad x_0 \le x \le x_0 + h$$

Hence

$$\left| \frac{dy}{dx} - \frac{d\tilde{y}}{dx} \right| \le |f(x, y) - f(x, \tilde{y})| + \epsilon$$

$$\le k|\tilde{y} - y| + \epsilon$$

by the Lipschitz condition; i.e.,

$$\left| \frac{dp}{dx} \right| \le k|p| + \epsilon, \qquad x_0 \le x \le x_0 + h \tag{5}$$

except for a finite number of points at which $[dp(x)]/dx$ fails to be defined.

Case I. Suppose $p(x) \ne 0$, $x_0 < x \le x_0 + h$. Hence, being continuous, it has the same sign, say without loss of generality $p(x) > 0$. Then a fortiori we can write (5) in the form

$$\frac{dp}{dx} \le kp(x) + \epsilon \tag{5'}$$

This may be written as

$$e^{-kx}[p'(x) - kp(x)] \leq \epsilon e^{-kx}$$

or, integrating from x_0 to x, where $x_0 \leq x \leq x_0 + h$

$$\int_{x_0}^{x} e^{-kx}[p'(x) - kp(x)]\, dx \leq \epsilon \int_{x_0}^{x} e^{-kx}\, dx \qquad (6)$$

The integrand on the left-hand side may have a finite number of simple discontinuities but it has a continuous indefinite integral. Hence we may write

$$[e^{-kx} p(x)]_{x_0}^{x} \leq \epsilon \left[-\frac{1}{k} e^{-kx} \right]_{x_0}^{x}$$

or

$$p(x) \leq e^{k(x-x_0)} p(x_0) + \frac{\epsilon}{k}[e^{k(x-x_0)} - 1] \qquad (7)$$

which is the required inequality.

Case II. If for all x, $p(x) = 0$, the theorem is obvious.

Case III. If $p(\bar{x}) \neq 0$ where $\bar{x}$ is some fixed number $x_0 \leq \bar{x} \leq x_0 + h$, but $p(x) = 0$ for some value of x, $x_0 \leq x < \bar{x}$; since $p(x)$ is continuous, there exists a number x_1, $x_0 \leq x_1 < \bar{x} \leq x_0 + h$ such that $p(x_1) = 0$, but $p(x) \neq 0$, $x_1 < x < \bar{x}$. Applying Case I to the interval (x_1, x) we have

$$p(\bar{x}) \leq e^{k(\bar{x}-x_1)} p(x_1) + \frac{\epsilon}{k}[e^{k(\bar{x}-x_1)} - 1]$$

$$= \frac{\epsilon}{k}[e^{k(\bar{x}-x_1)} - 1] \qquad (8)$$

which is an even stronger inequality than (4).

Hence the inequality (4) holds in all cases, since, if $p(x) < 0$, the same results follow by considering $|p(x)|$.

4. Uniqueness and Existence Theorems

Theorem 3. If $f(x, y)$ is continuous and satisfies a Lipschitz condition for y in a domain D, and if (x_0, y_0) is in D and $y(x)$ and $\bar{y}(x)$ are two exact solutions of $y' = f(x, y)$ in an interval $|x - x_0| \leq h$ such that $y(x_0) = \bar{y}(x_0) = y_0$, then $y(x) \equiv \bar{y}(x)$, $|x - x_0| \leq h$; i.e., there is at most one integral curve passing through any point of D.

Proof. Applying Theorem 2, we have $p(x_0) = y_0 - y_0 = 0$, and $\epsilon = \epsilon_1 + \epsilon_2 = 0$; hence $p = y - \bar{y} \equiv 0$.

We may state this in the form that under the above assumptions, two integral curves cannot meet or intersect at any point of D.

Observe that without the additional requirement of a Lipschitz condition uniqueness need not follow. For consider the differential equation

$$\frac{dy}{dx} = y^{\frac{1}{3}} \tag{1}$$

$f(x, y) = y^{\frac{1}{3}}$ is continuous at $(0, 0)$. But there are two solutions passing through $(0, 0)$, namely

(a) $$y \equiv 0$$

(b) $$\begin{cases} y = (\tfrac{2}{3}x)^{3/2} & x \geq 0 \\ y = 0 & x \leq 0 \end{cases} \tag{2}$$

Of course, $y^{\frac{1}{3}}$ does not satisfy the Lipschitz condition at $y = 0$. For, if $y_1 = \delta$ and $y_2 = -\delta$

$$\left| \frac{f(y_1) - f(y_2)}{y_1 - y_2} \right| = \frac{1}{\delta^{2/3}} \tag{3}$$

which is unbounded for δ arbitrarily small.

Theorem 4. *If $f(x, y)$ is continuous and satisfies the Lipschitz condition for y in a domain D, then for (x_0, y_0) in D there exists an exact solution of $y' = f(x, y)$ for $|x - x_0| \leq h$, where h is defined as in Theorem 1, such that*

$$y(x_0) = y_0$$

Proof. Given a monotone positive sequence $\{\epsilon_n\}$ approaching 0 as a limit, by Theorem 1 there exists a sequence $\{y_n(x)\}$ of functions satisfying

$$y_n'(x) = f(x, y_n(x)) \text{ up to } \epsilon_n, \qquad y_n(x_0) = y_0$$

over $|x - x_0| \leq h$; i.e.

$$|y_n'(x) - f(x, y_n(x))| \leq \epsilon_n, \qquad |x - x_0| \leq h \tag{4}$$

except for the finite set of points $x_i^{(n)}, i = 1, \cdots, m_n$.

Part I. The sequence $\{y_n(x)\}$ converges uniformly over $|x - x_0| \leq h$ to a (continuous) function $y(x)$.

For let n and p be positive integers; and apply Theorem 2 to $y_n(x)$, $y_{n+p}(x)$

$$|y_n(x) - y_{n+p}(x)| \leq \frac{\epsilon_n + \epsilon_{n+p}}{k} [e^{k(x - x_0)} - 1] \tag{5}$$

$$\leq \frac{2\epsilon_n}{k} [e^{kh} - 1]$$

whence the assertion.

Part II. The sequence $\left\{ \int_{x_0}^{x} f(t, y_n(t))\, dt \right\}$ approaches $\int_{x_0}^{x} f(t, y(t))\, dt$ uniformly for $|x - x_0| \leq h$.

Let R be the rectangle $|x - x_0| \leq h$, $|y - y_0| \leq Mh$, by hypothesis in *D*. Since $f(x, y)$ is continuous in the closed domain R, given $\epsilon > 0$, there exists $\delta > 0$ such that

$$|f(x, y_1) - f(x, y_2)| \leq \epsilon, \qquad (x, y_1), (x, y_2) \text{ in } R, \qquad |y_1 - y_2| \leq \delta \quad (6)$$

Likewise for this $\delta > 0$ there exists N such that

$$|y_n(x) - y(x)| \leq \delta, \qquad n > N, \qquad |x - x_0| \leq h \quad (7)$$

in virtue of Part I. Hence if $n > N$

$$|f(x, y_n(x)) - f(x, y(x))| \leq \epsilon, \qquad n > N, \qquad |x - x_0| \leq h \quad (8)$$

Therefore the sequence $\{f(x, y_n(x))\}$ converges uniformly to $f(x, y(x))$. Hence, by a well-known theorem we may reverse the order of integration and pass to the limit; which completes the proof of Part II.

Part III. $y(x)$ is differentiable, $y(x_0) = y_0$, and

$$y'(x) = f(x, y(x)), \qquad |x - x_0| \leq h$$

Integrating each side of (4) from x_0 to x_1 we have

$$\left| \int_{x_0}^{x} \left[\frac{dy_n(t)}{dt} - f(t, y_n(t)) \right] dt \right| \leq \epsilon_n |x - x_0| \leq \epsilon_n h \quad (9)$$

But since $y_n(t)$ is continuous

$$\left| y_n(x) - y_0 - \int_{x_0}^{x} f(t, y_n(t))\, dt \right| \leq \epsilon_n h \quad (10)$$

Approaching the limit by Part II, we have

$$y(x) - y_0 - \int_{x_0}^{x} f(t, y(t))\, dt = 0 \quad (11)$$

from which the assertion follows by an elementary theorem on functions defined by definite integrals.

Theorem 5. *Let $y(x)$ be an exact solution of $y' = f(x, y(x))$ under the assumptions of Theorem 4, and $\bar{y}(x)$ an approximate solution up to the error ϵ, for $|x - x_0| \leq h$, such that $\bar{y}(x_0) = y_0$. Then there exists a constant N independent of ϵ such that $|\bar{y}(x) - y(x)| \leq \epsilon N$, $|x - x_0| \leq h$.*

Proof. By Theorem 2

$$|\bar{y} - y| \leq \frac{\epsilon}{k}(e^{kh} - 1); \qquad \text{i.e., } N = \frac{e^{kh} - 1}{k} \quad (12)$$

Theorem 5 justifies our description of the function of Theorem 1 as an "approximate solution," for its actual value differs from that of the exact solution by a multiple of ϵ; whereas in our definition we only required it *to satisfy the differential equation* to within an error ϵ. To recapitulate our results: If $f(x, y)$ is continuous in D and satisfies a Lipschitz condition with constant k, and (x_0, y_0) is in D, then there exist quantities h and N such that

(a) There is a unique exact solution $y(x)$ of $y'(x) = f(x, y(x))$ passing through (x_0, y_0) and defined for $|x - x_0| \leq h$.

(b) There may be constructed by the Cauchy–Euler method approximate solutions $\tilde{y}(x)$ such that, if $\epsilon > 0$

$$|\tilde{y}'(x) - f(x, \tilde{y}(x))| \leq \epsilon, \qquad |x - x_0| \leq h, \qquad x \neq \xi_1, \xi_2, \cdots, \xi_n$$
$$|\tilde{y}(x) - y(x)| \leq N\epsilon, \qquad |x - x_0| \leq h \qquad\qquad (13)$$
$$\tilde{y}(x_0) = y(x_0) = y_0$$

Appendix to §4. Extension of the Existence Theorem

We have seen that without the Lipschitz condition on $f(x, y)$ the solution need not be unique. The condition is, however, superfluous for the proof of existence of a solution. This may be proved directly; we give a proof based upon Theorem 4 by the use of Weierstrass' polynomial approximation theorem.

Theorem 6. *If $f(x, y)$ is continuous in D, there is a solution of $y' = f(x, y)$ such that $y(x_0) = y_0$ where (x_0, y_0) lies in D, defined for $|x - x_0| \leq h = \min(a, b/M)$.*

Proof. We use the following two theorems of analysis:

Lemma 1 (Weierstrass). *If $f(x, y)$ is continuous in the region R, given $\epsilon > 0$, there exists a polynomial $P(x, y)$ such that*

$$|P(x, y) - f(x, y)| \leq \epsilon, \qquad (x, y) \text{ in } R$$

and

$$P(x, y) \leq f(x, y), \qquad (x, y) \text{ in } R$$

Definition 3. *A set S of functions $f(x)$ is equicontinuous over an interval $[a, b]$ if given $\epsilon > 0$ there exists $\delta > 0$ such that if x_1, x_2 in $[a, b]$, $|x_1 - x_2| \leq \delta$, then $|f(x_1) - f(x_2)| \leq \epsilon$, for all f in S.*

Lemma 2. *An infinite set S of functions uniformly bounded and equicontinuous over a closed interval $[a, b]$ contains a sequence converging uniformly over $[a, b]$.*

By virtue of Lemma 1, let $P_n(x, y)$ be a sequence of polynomials such that

$$|P_n(x, y) - f(x, y)| \leq \epsilon_n, \qquad |x - x_0| \leq h, \qquad |y - y_0| \leq Mh$$

where h and M are defined as usual, and the sequence $\{\epsilon_n\}$ approaches 0. Then $\{P_n(x, y)\}$ is uniformly bounded; and it may be assumed that M is the common bound of f and P_n. Since $P_n(x, y)$ satisfies Lipschitz conditions in y, by Theorem 4 there exists a sequence of functions $\{y_n(x)\}$ defined over $|x - x_0| \leq h$, satisfying $y_n{}'(x) = P_n\{x, y_n(x)\}$, and such that $y_n(x_0) = y_0$. By Theorem 4

$$|y_n(x) - y_n(x_0)| \leq Mh, \qquad |x - x_0| \leq h$$

i.e., $\{y_n(x)\}$ *is uniformly bounded over* $|x - x_0| \leq h$.

Furthermore, $\{y_n(x)\}$ *is equicontinuous over* $|x - x_0| \leq h$ for

$$y_n(x_2) - y_n(x_1) = y_n{}'(\xi)(x_2 - x_1), \qquad x_1 < \xi < x_2$$

for all x_1, x_2 in $[a, b]$. But

$$|y_n{}'(\xi)| = |P_n(\xi, y_n(\xi))| \leq M$$

Hence

$$|y_n(x_2) - y_n(x_1)| \leq |x_2 - x_1|M$$

Given $\epsilon > 0$, if $\delta = \epsilon/M$, then if $|x_2 - x_1| \leq \delta$; x_1, x_2 in $|x - x_0| \leq h$, then $|y_n(x_2) - y_n(x_1)| \leq \epsilon$ all n. Hence by Lemma 2, $\{y_n(x)\}$ has a subsequence converging uniformly over $|x - x_0| \leq h$ to a (continuous) function $y(x)$. It will do no harm if we denote the subsequence by the same symbol $\{y_n(x)\}$.

As in Theorem 4, Part II, we can prove that

$$\lim_{n \to \infty} \int_{x_0}^{x} f(t, y_n(t))\, dt \to \int_{x_0}^{x} f(t, y(t))\, dt$$

uniformly for $|x - x_0| \leq h$. By definition

$$y_n(x) - y_0 = \int_{x_0}^{x} P_n(t, y_n(t))\, dt$$

Hence we write

$$\left| y_n(x) - y_0 - \int_{x_0}^{x} f(t, y_n(t))\, dt \right| \leq \int_{x_0}^{x} |P_n(t, y_n(t)) - f(t, y_n(t))|\, dt$$

But if $n > N$

$$|P_n(x, y) - f(x, y)| \leq \epsilon_n, \qquad |x - x_0| \leq h, \qquad |y - y_0| \leq Mh$$

Hence for $n > N$

$$\left| y_n(x) - y_0 - \int_{x_0}^{x} f(t, y_n(t)) \right| |dt \leq \epsilon_n h$$

Letting $n \to \infty$ the above is valid with $y_n(x)$ replaced by $y(x)$ and so

$$\left| y(x) - y_0 - \int_{x_0}^{x} f(t, y(t))\, dt \right| = 0, \qquad |x - x_0| \leq h$$

since ϵ_n tends to 0. As in Theorem 4, this last formula proves that $y(x)$ actually is the required solution.

5. Solutions Containing Parameters

We consider next changes in the solution of a differential equation caused by either a small change in the initial conditions or a small variation in the function $f(x, y)$. Our general results will be that such small changes will bring about only small changes in the solution. This again will be of interest to the physicist, since it will assure him that small errors in the statement of a problem cannot change the answer too greatly.

We have seen already that the solution $y(x)$ of

$$y' = f(x, y), \qquad y(x_0) = y_0$$

contains the initial value y_0 as parameter. We determine first a minimum range of values of x and y_0 for which we can be assured that $y(x, y_0)$ will exist and be unique.

Theorem 7. *If $f(x, y)$ is continuous and satisfies a Lipschitz condition on y in $|x - x_0| \leq a$, $|y - \tilde{y}_0| \leq b$, $(x_0, \tilde{y}_0)$ some fixed point, there exists a unique solution $y(x, y_0)$ of*

$$y' = f(x, y), \qquad y(x_0) = y_0$$

in the region

$$|x - x_0| \leq h', \qquad |y_0 - \tilde{y}_0| \leq b/2, \qquad |y - y_0| \leq Mh' \qquad (1)$$

where $h' = min\,[a, (b/2M)]$, M being the bound of $f(x, y)$ in $|x - x_0| \leq a$, $|y - \tilde{y}_0| \leq b$.

Proof. The proof follows immediately upon applying Theorem 4 to the point (x_0, y_0), $|y_0 - \tilde{y}_0| \leq b/2$. Observe that the h' of this theorem is not less than half of the h for which the original existence theorem was proved.

Theorem 8. *If in some domain D, $f(x, y)$ is continuous and satisfies a Lipschitz condition on y for some k, and if a solution $y(x, y_0)$ of*

$$y' = f(x, y), \qquad y(x_0) = y_0$$

exists for some rectangle R, $|x - x_0| \leq h$, $|y_0 - \tilde{y}_0| \leq l$, then $y(x, y_0)$ is continuous in x and y_0 simultaneously in R.

Proof. Consider the solutions $y(x, y_0^{(1)})$, $y(x, y_0^{(2)})$, $y_0^{(1)}$, and $y_0^{(2)}$ satisfying $|y_0^{(i)} - \bar{y}_0| \leq l$, $i = 1, 2$. Applying Theorem 2 to these two functions over $|x - x_0| \leq h$, we obtain

$$|y(x, y_0^{(1)}) - y(x, y_0^{(2)})| \leq |y_0^{(1)} - y_0^{(2)}| e^{kh}, \qquad |x - x_0| \leq h \qquad (2)$$

This means that in R, $y(x, y_0)$ is continuous in y_0 *uniformly in x.* Since $y(x, y_0)$ is continuous in x for any value of y_0, it follows by a general theorem of real variables that in R, $y(x, y_0)$ is continuous in x and y_0 simultaneously.

Geometrically (2) means that two integral curves close enough together for *one value of x* stay close together for $|x - x_0| \leq h$.

With a little stronger restriction upon $f(x, y)$ we now prove a further property of $y(x, y_0)$, namely:

Theorem 9. *Under the assumptions of Theorem 8 and if in addition $(\partial f/\partial y)(x, y)$ exists in D and is continuous in x and y simultaneously, then $[\partial y(x, y_0)]/\partial y_0$ exists for (x, y_0) in R and is continuous in x and y_0 simultaneously.*

Proof. Let $\bar{y}_0$ be an arbitrary value of y_0 which is to remain fixed until the end of the argument, satisfying $|\bar{y}_0 - \tilde{y}_0| \leq l$. Let y_0 be a variable in the same interval. For convenience we write

$$y(x, \bar{y}_0) = \bar{y}(x), \qquad y(x, y_0) = y(x) \qquad (3)$$

We shall prove that the function

$$p(x, y_0) = \frac{y(x) - \bar{y}(x)}{y_0 - \bar{y}_0}, \qquad y_0 \neq \bar{y}_0 \qquad (4)$$

approaches a limit as $y_0 \to \bar{y}_0$, which limit must be $[\partial y(x, y_0)]/\partial y_0$ at $y_0 = \bar{y}_0$; and that this limit has the required properties.

In the first place we write

$$\frac{d}{dx} [y(x) - \bar{y}(x)] = f(x, y(x)) - f(x, \bar{y}(x))$$

or by the mean-value theorem

$$= [y(x) - \bar{y}(x)] \left[\frac{\partial}{\partial y} f(x, \bar{y}(x)) + \delta\{y(x), \bar{y}(x)\} \right] \qquad (5)$$

where as $|y(x) - \bar{y}(x)| \to 0$, $\delta\{y(x), \bar{y}(x)\} \to 0$. Since by Theorem 8

$$\lim_{y_0 \to \bar{y}_0} y(x) = \bar{y}(x) \text{ uniformly for } x, \qquad |x - x_0| \leq h$$

it follows that

$$\lim_{y_0 \to \bar{y}_0} \delta\{y(x), \bar{y}(x)\} = 0 \qquad (6)$$

uniformly in x.

By (4) and (5) we may write

$$\frac{\partial p(x, y_0)}{\partial x} = p(x, y_0) \left[\frac{\partial}{\partial y} f(x, \bar{y}(x)) + \delta\{y(x), \bar{y}(x)\} \right] \tag{7}$$

Since $p \neq 0$ for $y_0 \neq \bar{y}_0$ by the general uniqueness theorem, we may divide (7) by p and integrate explicitly with respect to x

$$p(x, y_0) = \exp\left\{ \int_{x_0}^{x} [f_y(x, \bar{y}(x)) + \delta\{y(x), \bar{y}(x)\}] \, dx \right\} \tag{8}$$

where the constant of integration is determined by the fact that $p(x_0, y_0) = 1$. Equation (8) is valid for all y_0. But $\lim\limits_{y_0 \to \bar{y}_0} p(x, y_0)$ exists; for by (6) and (8)

$$\frac{\partial y(x, \bar{y}_0)}{\partial y_0} = \lim_{y_0 \to \bar{y}_0} p(x, y_0) = \exp\left\{ \int_{x_0}^{x} f_y(x, \bar{y}(x)) \, dx \right\} \tag{9}$$

The right-hand side of (9) is continuous in x and $\bar{y}_0$ simultaneously; for $(\partial f/\partial y)$ (x, y) was continuous by hypothesis, and $\bar{y}(x) = y(x, \bar{y}_0)$ is continuous by Theorem 8, $\bar{y}_0$ now being considered the variable point. This completes the proof.

By similar (but considerably more complicated) reasoning we can prove under the same assumptions that if $f(x, y)$ in addition has a continuous first derivative with respect to some parameter μ in a certain interval, the solution $y(x)$ is also continuous and differentiable in μ. We omit the proof since this will appear as a special case of the general theory of Chapter 2.

We finally prove that the solution is only slightly changed if $f(x, y)$ is only slightly changed; precisely:

Theorem 10. *If in some domain D,*

(a) $F(x, y)$ and $f(x, y)$ are continuous,

(b) $f(x, y)$ satisfies a Lipschitz condition on y for some k,

(c) $|F(x, y) - f(x, y)| \leq \epsilon, \qquad (x, y) \text{ in } D, \tag{10}$

(d) for some point (x_0, y_0) in D, $y(x)$ and $\bar{y}(x)$ are admissible in $|x - x_0| \leq h$ and satisfy

$$y'(x) = f(x, y), \qquad \bar{y}'(x) = F(x, \bar{y}), \qquad y(x_0) = \bar{y}(x_0) = y_0$$

then

$$|y(x) - \bar{y}(x)| \leq \frac{\epsilon}{k}(e^{kh} - 1), \qquad |x - x_0| \leq h \tag{11}$$

Proof. By (10), $\bar{y}(x)$ is an approximate solution of $y' = f(x, y)$ with error ϵ; applying Theorem 5 we immediately obtain (11).

For example, in the neighborhood of the origin we can replace $y' = \sin(xy)$, which cannot be integrated explicitly, by $y' = xy$, which can; and for sufficiently small values of x and y the error will be arbitrarily small.

PART B.
CONTINUATION OF SOLUTIONS.   OTHER METHODS

6. Continuation of Solutions

In Part A we proved essentially the following: If in a domain D the function $f(x, y)$ is continuous and satisfies a Lipschitz condition in y uniformly, for every point (x_0, y_0) of D there is a rectangle R_0 such that the integral curve $y(x)$ of $y' = f(x, y)$ passing through (x_0, y_0) can be extended at least to the right and left sides of R_0. Since R_0 lies in D, by applying Theorem 4 to the point at which the integral curve goes out of R_0, we can extend the region in which the curve is defined. We now prove that if D is bounded, the integral curve passing through any point of D may be extended *up to the boundary of D*.

We remark first that since D is bounded, and the integral curve cannot pass out of D, there exist positive numbers l and m, such that the integral curve passing through (x_0, y_0) can be defined in the open interval

$$x_0 - l < x < x_0 + m \tag{1}$$

but not outside it. Our theorem will demonstrate that as $x \to x_0 + m$ the integral curve approaches the boundary of D; a similar proof holds for approach on the left.

Theorem 11. Let D be an arbitrary bounded domain, and let $f(x, y)$ be continuous in D and satisfy a Lipschitz condition in y locally in D; i.e., for every point of D there must be a neighborhood N in which the condition is satisfied uniformly. [For example, it is sufficient that $(\partial/\partial y)f(x, y)$ exist and be continuous in D.] Let the solution $y(x)$ of $y' = f(x, y)$ passing through (x_0, y_0) be defined on the right only for

$$x_0 \leq x < x_0 + m \tag{1'}$$

Then if $p(x)$ is the distance of the point P, $(x, y(x))$ from the boundary C of D

$$\lim_{x \to x_0 + m} p(x) = 0 \tag{2}$$

Proof. Let $\epsilon > 0$ and let S be the region in D consisting of all points of D having a distance greater than ϵ from C. Suppose the solution $(x, y(x))$ has points in S as $x \to x_0 + m$. Then there exists a monotone increasing sequence $\{x_j\}$, $j = 1, 2, \cdots$, such that $x_j \to x_0 + m$ and the points P_j, $(x_j, y(x_j))$ are all in S. Since S is bounded, the points P_j have at least one cluster point $\tilde{P}$, $(\tilde{x}, \tilde{y})$, which is in S or on its boundary and hence is inside of D. Since $x_j \to x_0 + m$, $\tilde{x} = x_0 + m$ so that $x_j < \tilde{x}$ for all j.

Since $\tilde{P}$ is in S, $\tilde{P}$ is the center of a rectangle R which is in D and with edges $x = \tilde{x} \pm a$, $y = \tilde{y} \pm b$ for some positive a and b. Let M be a bound on $|f|$ in R. Then by decreasing a if necessary we can assume that $a < b/(4M)$.

Let R_1 be the rectangle with edges $x = \tilde{x} - a$, $x = \tilde{x}$, and $y = \tilde{y} \pm \frac{1}{2}b$. All points P_j must be in R_1 for j large enough. In particular, then, for some n, P_n, $(x_n, y(x_n))$ is in R_1. Hence from Theorem 4 it follows that the solution of $y' = f(x, y)$ through P_n must exist to the right of x_n and stay in R. Indeed for the case here, h of Theorem 4 is given by

$$h = \min \left(\tilde{x} + a - x_n, \tfrac{1}{2}b/M \right)$$

and since $\tilde{x} + a - x_n \leq 2a$ and $\frac{1}{2}b/M > 2a$, it follows that $h = \tilde{x} + a - x_n$. Hence the solution $y(x)$ exists for $x_n \leq x \leq \tilde{x} + a$. Since $\tilde{x} = x_0 + m$, this contradicts the fact that the solution $y(x)$ is in D only for $x_0 - l < x < x_0 + m$. Hence as $x \to x_0 + m$, the solution can have no points in S. Since ϵ is arbitrary, this proves the theorem.

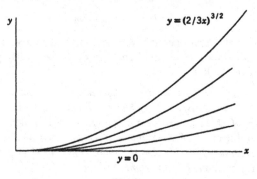

Fig. 2

This does not necessarily mean that the integral curves will actually approach some particular point of C; they might oscillate over a finite interval in the neighborhood of the boundary. Also, a number of integral curves may approach the same point of the boundary; e.g., if D is any bounded domain such that in D, $x > 0$, and $f(x, y) = y^{\frac{1}{3}}$, then any integral curve lying between $y = (\frac{2}{3}x)^{\frac{3}{2}}$ and $y = 0$ will approach the origin on the left (Fig. 2).

If D is unbounded, Theorem 11 does not hold as stated; however, we may replace it by a corollary:

Corollary. *If D is unbounded, under the other conditions of Theorem* 11, *as $x \to x_0 + m$ either*

(a) *$y(x)$ becomes unbounded, or*
(b) *$y(x)$ actually approaches the boundary of D.*

For if $y(x)$ does not become unbounded, we may replace D by a suitable bounded domain, and Theorem 11 is applicable.

Under certain conditions we can go farther and state that, even though D is unbounded, the integral curves will approach a real boundary.   For example:

Theorem 12. Let $f(x, y)$ be continuous in D; $x_1 < x < x_2$, $-\infty < y < \infty$, and satisfy a Lipschitz condition on y uniformly in D. Then the integral curve $y(x)$ passing through any point (x_0, y_0) of D is defined over the whole open interval $x_1 < x < x_2$.

Proof. Suppose there is a point $\bar{x}$, $x_0 < \bar{x} < x_2$, beyond which $y(x)$ cannot be continued.   Since the Lipschitz condition is satisfied uniformly for some k, we may apply Theorem 2 to the approximate solution $\bar{y}(x) \equiv y_0$ over the interval $x_0 \leq x \leq \bar{x} - \epsilon$, for any $\epsilon > 0$.   The error of $\bar{y}(x)$ is

$$\max \left| \frac{d\bar{y}}{dx} - f(x, \bar{y}) \right| \equiv \max |f(x, y_0)|, \qquad x_0 \leq x \leq \bar{x} \tag{3}$$

This maximum is finite, since $f(x, y_0)$ is continuous over the closed interval $x_0 \leq x \leq \bar{x} < x_2$; let it be M.   Then by Theorem 2

$$|y(x) - y_0| \leq \frac{M}{k} [e^{k(x - x_0)} - 1], \qquad x_0 \leq x \leq \bar{x} - \epsilon \tag{4}$$

Hence $y(x)$ is *bounded* in the neighborhood of $x = \bar{x}$.   Therefore by the Corollary to Theorem 11, $(x, y(x))$ must approach the boundary of D as $x \to \bar{x}$.   This is manifestly impossible, and therefore $y(x)$ must be defined for $x_0 \leq x \leq x_2$.   Similarly, $y(x)$ must be defined for $x_1 < x \leq x_0$.   Hence the theorem.

For example, a solution of the equation

$$\frac{dy}{dx} = P(x) \cos y + Q(x) \sin y$$

[where $P(x)$ and $Q(x)$ are polynomials] passing through any given point must be defined *for all values of x.*

Now that we have shown the existence and uniqueness of solutions in very general domains, we prove that the Cauchy–Euler method may be used to approximate an exact solution *arbitrarily closely* between *any two values of x* for which the solution is defined.

Theorem 13. *Under the hypotheses of Theorem 11, let $y(x)$ be an exact solution of $y' = f(x, y)$ in some domain D, passing through (x_0, y_0) and defined for $x = x_1$. Then the value of $y(x_1)$ may be determined by the Cauchy–Euler method to any degree of accuracy by a sufficiently fine subdivision of the interval $x_0 \leq x \leq x_1$.*

Proof. The boundary C of D and the curve $y(x)$, $x_0 \leq x \leq x_1$, have a nonzero distance d, being both closed, disjoint, bounded sets. Let η be any number, $0 < \eta < d$. The strip S, $x_0 \leq x \leq x_1$, $|y - y(x)| \leq \eta$, lies entirely in D. Hence there is some value k of the Lipschitz constant for $f(x, y)$, valid throughout S. We prove that if by the Cauchy–Euler method we begin to construct an approximate solution $\tilde{y}(x)$ at (x_0, y_0), satisfying the differential equation with an error ϵ

$$\epsilon = \frac{k\eta}{e^{k|x_1 - x_0|} - 1} \tag{5}$$

then $\tilde{y}(x)$ will be defined over the whole interval $x_0 \leq x \leq x_1$, and $|\tilde{y}(x_1) - y(x_1)| \leq \eta$.

We know by Theorem 2 that $\tilde{y}(x)$ will certainly be defined in some interval to the right of x_0, say $x_0 \leq x \leq \bar{x}$. Then by Theorem 5 we can compute the actual error; it is

$$|\tilde{y}(\bar{x}) - y(\bar{x})| \leq \eta \frac{e^{k|\bar{x} - x_0|} - 1}{e^{k|x_1 - x_0|} - 1} \tag{6}$$

Hence if $\bar{x} < x_1$, $|\tilde{y}(\bar{x}) - y(\bar{x})| < \eta$; and therefore $\tilde{y}(x)$ *still is in the strip S*. As long as $\tilde{y}(x)$ stays in S, we can continue the approximating process, since in S, $f(x, y)$ satisfies the conditions of Theorem 2. Hence $\tilde{y}(x)$ is defined up to $x = x_1$, and from (6) $|\tilde{y}(x_1) - y(x_1)| \leq \eta$; whence the theorem.

7. Other Methods of Solution

Another method of proving the local existence of a solution of $y' = f(x, y)$ is Picard's *method of successive approximations*. The idea behind this method is as follows. Suppose $y = \phi(x)$ is an approximate solution of

$$\frac{dy}{dx} = f(x, y), \qquad y(x_0) = y_0 \tag{1}$$

with the initial condition $\phi(x_0) = y_0$. Let us replace the unknown function y in the right-hand side of (1) by the function $\phi(x)$. This yields the differential equation

$$\frac{dy}{dx} = f(x, \phi(x)), \qquad y(x_0) = y_0 \tag{2}$$

which can be solved by a quadrature. It seems very plausible that the solution of (2) will better approximate the exact solution of (1) than will $\phi(x)$. Picard's existence proof consists of repeating this process to form an infinite sequence of approximate solutions which are then shown to converge to an exact solution. The process also produces approximate solutions of any required degree of accuracy. Observe the analogy between this method and Newton's method of approximating solutions of algebraic equations.

It will be more convenient to transform (1) into an *integral equation*; we state formally:

Lemma 1. *Let $f(x, y)$ be continuous over a domain D, and let $y(x)$ be any continuous function defined over some interval $x_1 \leq x \leq x_2$, and admissible [i.e., such that $(x, y(x))$ is in D]. Let x_0 be any point such that $x_1 < x_0 < x_2$. Then a necessary and sufficient condition that $y(x)$ be a solution of the differential equation (1) is that it be a solution of the integral equation*

$$y(x) = y_0 + \int_{x_0}^{x} f(t, y(t)) \, dt, \qquad x_1 \leq x \leq x_2 \tag{3}$$

The proof is trivial; we have already used the lemma implicitly.

We shall demonstrate the existence of the solution of (3) by Picard's method under the same assumptions as in Theorem 4. This theorem is closely related to the so-called *fixed point theorems* of topology, which state that given a continuous transformation of an abstract space S into itself, under certain conditions, there must exist points which remain unchanged under the transformation. Now we may take as the "points" of S a set of continuous functions ϕ over $x_1 \leq x \leq x_2$. If we consider the transformation $L(\phi)$

$$L(\phi) = y_0 + \int_{x_0}^{x} f(t, \phi(t)) \, dt \tag{4}$$

it can be shown that S and L satisfy the conditions of a certain very general fixed-point theorem, and hence that there exists at least one "point" ϕ for which $L(\phi) = \phi$, i.e., for which (3) holds.

We prove our theorem, however, by more elementary and constructive means.

Lemma 2. Let R: $|x - x_0| \leq h$, $|y - y_0| \leq Mh$ be the rectangle of Theorem 2 in which $f(x, y)$ is continuous and bounded by M. Let $\phi_0(x)$ be an arbitrary function with continuous first derivative and admissible in R [e.g., set $\phi_0(x) \equiv y_0$]. Define $\phi_1(x)$ by

$$\phi_1(x) = y_0 + \int_{x_0}^{x} f(t, \phi_0(t)) \, dt, \qquad |x - x_0| \leq h \tag{5}$$

Then $\phi_1(x)$ is admissible and $\phi_1(x_0) = y_0$.

Proof. Obviously

$$|\phi_1(x) - y_0| \leq \int_{x_0}^{x} |f(t, \phi_0(t))| \, dt \leq \max |f(t, \phi_0(t))| |x - x_0|$$

$$\leq Mh, \quad |x - x_0| \leq h \tag{6}$$

Hence $\phi_1(x)$ is admissible.

By repeated applications of the lemma we may define an infinite sequence of admissible functions by the recursion formula

$$\phi_{n+1}(x) = y_0 + \int_{x_0}^{x} f(t, \phi_n(t)) \, dt, \qquad n = 0, 1, 2, \cdots, \qquad |x - x_0| \leq h$$

We now state our main result: $\tag{7}$

Theorem 14. If in addition to the assumption of Lemma 2, $f(x, y)$ satisfies a Lipschitz condition on y in R for some constant k, then the sequence $\{\phi_n(x)\}$ converges uniformly and absolutely to a function satisfying the integral equation (3) and hence the differential equation (1).

Proof. Let D be the error with which $\phi_0(x)$ satisfies (1) in $|x - x_0| \leq h$

$$\left| \frac{d\phi_0(x)}{dx} - f(x, \phi_0(x)) \right| \leq D, \qquad |x - x_0| \leq h \tag{8}$$

Integrating (8) from x_0 to x, we obtain

$$|\phi_0(x) - \phi_1(x)| = \left| \phi_0(x) - y_0 - \int_{x_0}^{x} f(t, \phi_0(t)) \, dt \right| \leq D|x - x_0| \tag{9}$$

We wish to prove

$$|\phi_n(x) - \phi_{n-1}(x)| \leq \frac{k^{n-1} D |x - x_0|^n}{n!}, \qquad |x - x_0| \leq h \tag{10}$$

We give a proof by induction for $x \geq x_0$. Equation (10) is certainly true for $n = 1$. Assume (10) for $n = m$. By the definition of $\phi_n(x)$

$$|\phi_{m+1}(x) - \phi_m(x)| \leq \int_{x_0}^{x} |f(t, \phi_m(t)) - f(t, \phi_{m-1}(t))| \, dt$$

and by the Lipschitz condition

$$\leq k \int_{x_0}^{x} |\phi_m(t) - \phi_{m-1}(t)| \, dt \tag{11}$$

Since we have assumed (10) true for $n = m$, we have

$$|\phi_{m+1}(x) - \phi_m(x)| \leq \frac{k^m D}{m!} \int_{x_0}^{x} (t - t_0)^m \, dt \tag{12}$$

whence

$$|\phi_{m+1}(x) - \phi_m(x)| \leq \frac{k^m D |x - x_0|^{m+1}}{(m+1)!} \tag{13}$$

This being (10) written for $n = m + 1$, the proof by induction is complete.
Equation (10) may now be written

$$|\phi_n(x) - \phi_{n-1}(x)| \leq \frac{k^{n-1} D h^n}{n!}, \qquad |x - x_0| \leq h \tag{14}$$

This proves that the sequence $\{\phi_n(x)\}$, or equivalently the series

$$\phi_0(x) + \sum_{n=1}^{\infty} (\phi_n(x) - \phi_{n-1}(x)) \tag{15}$$

converges uniformly to a continuous function $\phi(x)$. But in (7) we may
pass to the limit and interchange the limit process and integration process,
as we have done several times before. Hence

$$\phi(x) = \lim_{n \to \infty} \phi_n(x) = y_0 + \lim_{n \to \infty} \int_{x_0}^{x} f(t, \phi_n(t)) \, dt$$

$$= y_0 + \int_{x_0}^{x} f(t, \phi(t)) \, dt \tag{16}$$

This completes the proof.

Observe that the existence of a solution has been proved only for the
interval $|x - x_0| \leq h$. Thus Picard's method (unlike the Cauchy-Euler
method) is suitable for finding approximate solutions only in a *small*
interval about x_0. But in such an interval it is very convenient. The
error after the nth approximation may be easily computed; it is

$$|\phi(x) - \phi_n(x)| \leq \sum_{r=n}^{\infty} |\phi_{r+1}(x) - \phi_r(x)| \leq \frac{D}{k} \sum_{r=n+1}^{\infty} \frac{(kh)^r}{r!}$$

$$\leq \frac{D}{k} \frac{(kh)^n}{n!} \sum_{r=1}^{\infty} \frac{(kh)^r}{r!} = \frac{D}{k} (e^{kh} - 1) \frac{(kh)^n}{n!} \tag{17}$$

Finally we remark that the Cauchy-Euler method may be generalized by using more refined types of approximation to the exact solution. For example, the method of Adams and Störmer uses parabolic arcs whose extra constants are so chosen as to fit more closely the part of the curve already constructed. Runge's method uses line segments whose slope is determined by the value of $f(x, y)$ at several points, using an interpolation formula. These methods supplant our crude ones in the solution of any practical problem.

2

Systems of Differential Equations

1. Introduction. Vector Notation

We now consider a *system of differential equations* of the form:

$$\frac{dx_1}{dt} = f_1(x_1, x_2, \cdots, x_n; t)$$

$$\frac{dx_2}{dt} = f_2(x_1, x_2, \cdots, x_n; t)$$

$$\quad \cdot \qquad \cdot$$
$$\quad \cdot \qquad \cdot \tag{1}$$
$$\quad \cdot \qquad \cdot$$

$$\frac{dx_n}{dt} = f_n(x_1, x_2, \cdots, x_n; t)$$

where $f_1, f_2, \cdots, f_n$ are single-valued functions continuous in a certain domain of their arguments, and $x_1, x_2, \cdots, x_n$ are unknown functions of the real variable t. A *solution* of (1) is a set of functions $x_1(t), x_2(t), \cdots, x_n(t)$ satisfying (1) for some interval $t_1 < t < t_2$.

To treat the system (1) it is convenient to introduce vector notation. An *n-dimensional vector* X is defined as an ordered set of n real numbers, $x_1, x_2, \cdots, x_n$, called the *components* of X, and X is designated by

$$X = (x_1, x_2, \cdots, x_n)$$

23

We shall always use capital letters for vectors and small letters for real variables (scalars). We assume that the reader is familiar with the elements of vector theory. By $|X|$ we denote the *norm* of the vector X

$$|X| = (x_1^2 + x_2^2 + \cdots + x_n^2)^{\frac{1}{2}}$$

If X is a function of a real variable t

$$X(t) = \{x_1(t), \cdots, x_n(t)\}$$

it is *continuous in* t if and only if each of its components is continuous in t. If each of its components is differentiable, $X(t)$ has the *derivative*

$$\frac{dX(t)}{dt} = \left\{ \frac{dx_1(t)}{dt}, \frac{dx_2(t)}{dt}, \cdots, \frac{dx_n(t)}{dt} \right\}$$

If X is a function of the components $y_1, y_2, \cdots, y_n$ of another vector Y, it is *continuous in* Y if and only if it is continuous in the n variables $y_1, y_2, \cdots, y_n$ simultaneously.

Let us therefore in terms of the system (1) define the vectors

$$X = (x_1, x_2, \cdots, x_n), \qquad F(X, t) = \{f_1, f_2, \cdots, f_n\} \tag{2}$$

where by assumption F is continuous in X and t simultaneously. Then the system (1) may be written as the single *vector equation*

$$\frac{dX}{dt} = F(X, t) \tag{3}$$

The vector equation (3) is obviously analogous to the scalar equation of Chapter 1

$$\frac{dx}{dt} = f(x, t) \tag{4}$$

and is in fact equivalent to it when $n = 1$. We shall discover that all the theorems of Chapter 1 may be generalized so as to hold for the vector equation (3).

We shall also at times consider the set of n numbers $(x_1, x_2, \cdots, x_n)$ as defining a *point* in n-dimensional Euclidean space. Then of course every such point defines a vector and vice versa. We shall find it convenient to use the two ideas of vector and point interchangeably; e.g., we shall define the range in which the vector X may vary by saying that the point $(x_1, x_2, \cdots, x_n)$ (or even "the point X") may vary through a certain region R of n-dimensional space. We shall use vector notation wherever algebraic processes are involved. This double terminology will lead to no confusion.

In §§ 2–5 of this chapter the theory will closely parallel that of Chapter 1, so we shall give detailed proofs only where there is a real difference from the simpler case. In these sections the theorems are numbered so as to correspond with Chapter 1.

2. Approximate Solutions

Let $X_0 = (x_{10}, x_{20}, \cdots, x_{n0})$ and t_0 be fixed. Let $F(X, t)$ be continuous in the $(n + 1)$-dimensional region R

$$|X - X_0| \leq b, \qquad |t - t_0| \leq a \tag{1}$$

Since $F(X, t)$ is continuous, $|F|$ is bounded by some number M in R. Then as before we define an *approximate solution of*

$$\frac{dX}{dt} = F(X, t) \tag{2}$$

with error ϵ as a continuous admissible function $X(t)$ with piecewise continuous derivative, such that

$$X(t_0) = X_0 \tag{3}$$

and satisfying

$$|X'(t) - F(X(t), t)| \leq \epsilon \tag{4}$$

for all points of some interval at which the left-hand side is defined. As before, we have immediately:

Theorem 1. Given $\epsilon > 0$, *there can be constructed an approximate solution of* (2) *with error* ϵ *over the interval*

$$|t - t_0| \leq h = \min\left(a, \frac{b}{M}\right) \tag{5}$$

For exactly as before, there exists $\delta > 0$ such that

$$|F(X_1, t_1) - F(X_2, t_2)| \leq \epsilon$$

for $(X_1, t_1), (X_2, t_2)$ in R, $|X_1 - X_2| \leq \delta, |t_1 - t_2| \leq \delta$. Choose numbers $t_i, i = 1, 2, \cdots, m - 1$ such that

$$t_0 < t_1 < t_2 < \cdots < t_m = t_0 + h$$

$$|t_i - t_{i-1}| \leq \min\left(\delta, \frac{\delta}{M}\right), \qquad i = 1, 2, \cdots, m$$

Then the approximate solution will be given by the recursion formula

$$X(t) = X_{i-1} + (t - t_{i-1})F(X_{i-1}, t_{i-1})$$

for $t_{i-1} \leq t \leq t_i$, where $X_i = X(t_i)$.

The details of the proof proceed as before.

3. The Lipschitz Conditions

We shall next prove the fundamental inequality. The necessary extension of the Lipschitz condition is given by the following definitions and lemmas.

Definition 1. A continuous scalar function $f(X, t)$ defined in some $(n + 1)$-dimensional region R is said to satisfy a Lipschitz condition on X for k if

$$|f(X_1, t) - f(X_2, t)| \leq k|X_1 - X_2| \tag{1}$$

for (X_1, t), (X_2, t) in R.

Lemma 1. Let R be convex in the components $x_1, x_2, \cdots, x_n$ of X; i.e., let the n-dimensional region defined by (X, t) in R, $t = t_0$, be convex for each t_0. Then if each of the partial derivatives $[\partial f(X, t)]/\partial x_i$, $i = 1, 2, \cdots, n$, exists and is bounded by N in R, $f(X, t)$ satisfies a Lipschitz condition on X in R with constant $k = nN$.

Lemma 2. Let R be imbedded in a larger region D and let every point of R have a distance from the boundary of $D \geq \delta$. Then if $f(X, t)$ and $[\partial f(X, t)]/\partial x_i$, $i = 1, 2, \cdots, n$, are bounded by M and N respectively in $D, f(X, t)$ satisfies a Lipschitz condition on X in R for

$$k = max \left(\frac{2M}{\delta}, nN \right) \tag{2}$$

Proofs. These are analogues of the lemmas of Chapter 1, § 3, the only difference being in the value of k. By the mean-value theorem for several variables

$$|f(x_1, x_2, \cdots, x_n; t) - f(x_1', x_2', \cdots, x_n'; t)|$$

$$\leq \sum_{i=1}^{n} \left| (x_i - x_i') \frac{\partial f}{\partial x_i} \right| \leq N \sum_{i=1}^{n} |x_i - x_i'| \tag{3}$$

the partial derivative being evaluated at some point of R (for Lemma 1) or of D (for Lemma 2). But

$$\sum_{i=1}^{n} |x_i - x_i'| \leq n|X - X'| \tag{4}$$

whence the proofs proceed as before.

Definition 2. A vector function $F(X, t)$ is said to satisfy a Lipschitz condition on X for constant k if

$$|F(X_1, t) - F(X_2, t)| \leq k|X_1 - X_2| \tag{5}$$

Lemma 3. *A vector function* $F(X, t)$ *satisfies a Lipschitz condition on* X *if and only if each of its components* $f_i(X, t)$ *does (where the constants may be different).*

Proof. (1) If k holds for $F(X, t)$, then

$$|f_i(X_1, t) - f_i(X_2, t)| \leq |F(X_1, t) - F(X_2, t)| \leq k|X_1 - X_2| \qquad (6)$$

(2) If k holds for each f_i, then

$$|F(X_1, t) - F(X_2, t)| = \left[\sum_{i=1}^{n} \{f_i(X_1, t) - f_i(X_2, t)\}^2 \right]^{\frac{1}{2}}$$

$$\leq \sum_{i=1}^{n} |f_i(X_1, t) - f_i(X_2, t)| \leq nk|X_1 - X_2| \qquad (7)$$

Finally we shall need the following:

Lemma 4. *If a vector function* $X(t)$ *is differentiable over some interval* $t_1 < t < t_2$, *then wherever* $|X(t)| \neq 0$, $[d|X(t)|]/dt$ *exists and*

$$\left| \frac{d|X(t)|}{dt} \right| \leq \left| \frac{dX(t)}{dt} \right| \qquad (8)$$

Proof. If $|X(t)| \neq 0$, since $|X(t)| = \left[\sum_{i=1}^{n} x_i^2(t) \right]^{\frac{1}{2}}$, it follows that

$$\frac{d|X(t)|}{dt} = \frac{\sum_{i=1}^{n} x_i \dfrac{dx_i}{dt}}{|X(t)|}$$

Furthermore, in any case

$$\left| \frac{|X(t + \Delta t)| - |X(t)|}{\Delta t} \right| \leq \left| \frac{X(t + \Delta t) - X(t)}{\Delta t} \right| \qquad (9)$$

and from the existence of both derivatives it is clear that the same inequality will hold in the limit. Hence the lemma.

4. The Fundamental Inequality

We can now state the fundamental inequality for the vector equation (2) of § 2:

Theorem 2. *Let* $F(X, t)$ *be continuous in some* $(n + 1)$-*dimensional region* R *and satisfy a Lipschitz condition on* X *with constant* k. *Let* $X_1(t)$ *and* $X_2(t)$ *be two approximate solutions of* (2) *of* § 2 *over some interval* $|t - t_0| \leq b$ *with errors* ϵ_1 *and* ϵ_2 *respectively. Set*

$$P(t) = X_1(t) - X_2(t); \qquad p(t) = |P(t)|; \qquad \epsilon = \epsilon_1 + \epsilon_2 \qquad (1)$$

Then

$$|p(t)| \leq e^{k|t-t_0|} |p(t_0)| + \frac{\epsilon}{k} (e^{k|t-t_0|} - 1) \tag{2}$$

for $|t - t_0| \leq b$.

Proof. Wherever $P'(t)$ exists and $P(t) \neq 0$, by Lemma 4 $p'(t)$ exists and

$$|p'(t)| \leq |P'(t)|$$

By the definition of approximate solution

$$|P'(t)| = |X_1'(t) - X_2'(t)| \leq |F(X_1(t), t) - F(X_2(t), t)| + \epsilon$$

and by the Lipschitz condition on F

$$|P'(t)| \leq k|X_1(t) - X_2(t)| + \epsilon = kp(t) + \epsilon$$

whence a fortiori we have

$$p'(t) \leq kp(t) + \epsilon \tag{3}$$

From here on the proof follows exactly as in Chapter 1, since we need to integrate (3) only over intervals in which $p(t) \neq 0$, and therefore $p'(t)$ is defined except for a finite number of points. There is the additional simplification that in the present case $p(t)$ is nonnegative.

5. Existence and Properties of Solutions of the System

From the fundamental inequality our basic results follow exactly as in Chapter 1. We omit the proofs of the following theorems; they differ only formally from those of the simpler case.

Theorem 3 (Uniqueness). Let $F(X, t)$ be continuous and satisfy a Lipschitz condition on X in some neighborhood of the fixed point (X_0, t_0). Then there can exist at most one solution of

$$\frac{dX}{dt} = F(X, t) \tag{1}$$

such that $X(t_0) = X_0$.

Theorem 4 (Existence). Let $F(X, t)$ be continuous and satisfy a Lipschitz condition on X in the $(n + 1)$-dimensional region R

$$|X - X_0| \leq b, \qquad |t - t_0| \leq a \tag{2}$$

Let M be the upper bound of $|F|$ in R. Then there exists a solution $X(t)$ of (1) defined over the interval

$$|t - t_0| \leq h = \min \left(a, \frac{b}{M} \right) \tag{3}$$

(We omit the analogues of Theorems 5 and 6.)

Theorem 7. *Let* $F(X, t)$ *satisfy the conditions of Theorem* 4. *Then there exists a region* S

$$|X - \tilde{X}_0| \le a' < a, \qquad |t - t_0| \le b' < b \tag{4}$$

with the following property; if X_0 *is a vector such that* $|X_0 - \tilde{X}_0| \le a'$, *then there exists a unique solution* $X(t)$ *of* (1) *with the initial value* $X(t_0) = X_0$, *defined for* $|t - t_0| \le b'$.

Theorem 8. *Let* S *be any region with the properties of Theorem* 7. *Then the solution* $X(t)$ *of* (1), *considered as a function of its initial value* $X_0 = X(t_0)$, *is continuous in* X_0 *and* t *simultaneously, for* (X_0, t) *in* S.

The proof of the *differentiability* of the solutions with respect to their initial values is more difficult; we give it in full.

Theorem 9. *Let* $F(X, t)$ *satisfy the conditions of Theorem* 4 *for* R

$$|X - \tilde{X}_0| \le a, \qquad |t - t_0| \le b \tag{5}$$

Write

$$F(X, t) = \{f_1(X, t), f_2(X, t), \cdots, f_n(X, t)\}$$
$$X(X_0, t) = \{x_1(X_0, t), x_2(X_0, t), \cdots, x_n(X_0, t)\} \tag{6}$$
$$X_0 = \{x_{10}, x_{20}, \cdots, x_{n0}\}$$

where $\dfrac{dX(t)}{dt} = F(X, t)$, $X(t_0) = X_0$. *Let* $\dfrac{\partial}{\partial x_j} f_i(X, t)$; $i, j = 1, 2, \cdots, n$, *exist and be continuous in* X *and* t *simultaneously,* (X, t) *in* R. *Then* $\dfrac{\partial x_i(X_0, t)}{\partial x_{j0}}$; $i, j = 1, 2, \cdots, n$, *exist and are continuous in* X_0 *and* t *simultaneously,* (X_0, t) *in* S, *where* S *is the region of Theorem* 7. *Hence* $\dfrac{\partial X(X_0, t)}{\partial x_{j0}}$; $j = 1, 2, \cdots, n$, *exist and have the same properties.*

Proof. It suffices to prove that $\dfrac{\partial x_i(X_0, t)}{\partial x_{10}}$ exists and has the required properties. Let X_0 be an arbitrary fixed vector such that (X_0, t_0) is in S. Let Δx_{10} be some number sufficiently small so that the solutions $(x_1, x_2, \cdots, x_n)$ and $(x_1 + \Delta x_1, x_2 + \Delta x_2, \cdots, x_n + \Delta x_n)$ of (1) with initial values $(x_{10}, x_{20}, \cdots, x_{n0})$ and $(x_{10} + \Delta x_{10}, x_{20}, \cdots, x_{n0})$ are defined for $|t - t_0| \le b'$. This may always be done. Set

$$p_i(\Delta x_{10}, t) = \frac{\Delta x_i(t)}{\Delta x_{10}}, \qquad i = 1, 2, \cdots, n \tag{7}$$

where obviously

$$p_i(\Delta x_{10}, t_0) = \begin{cases} 1, & i = 1 \\ 0, & i \ne 1 \end{cases}$$

We shall prove that

$$\lim_{\Delta x_{10} \to 0} p_i(\Delta x_{10}, t), \qquad i = 1, 2, \cdots, n \tag{8}$$

exists *uniformly* for (X_0, t) in S. By Theorem 8, $x_i(t)$ [hence $p_i(\Delta x_{10}, t)$] is continuous in $x_{10}, x_{20}, \cdots, x_{n0}, t$ simultaneously for (X_0, t) in S. Therefore by virtue of the uniform convergence of (8) it will follow that

$$\frac{\partial x_i(X_0, t)}{\partial x_{10}} = \lim_{\Delta x_{10} \to 0} p_i(\Delta x_{10}, t)$$

has the required properties.

By applying the fundamental inequality to the two solutions $(x_1, x_2, \cdots, x_n)$ and $(x_1 + \Delta x_1, x_2 + \Delta x_2, \cdots, x_n + \Delta x_n)$, we obtain

$$\sqrt{\Delta x_1{}^2 + \Delta x_2{}^2 + \cdots + \Delta x_n{}^2} \le e^{kb}|\Delta x_{10}|, \qquad |t - t_0| \le b' \tag{9}$$

By the theorem of the mean in n variables, we may write

$$\frac{d\Delta x_i(t)}{dt} = f_i(x_1 + \Delta x_1, x_2 + \Delta x_2, \cdots, x_n + \Delta x_n; t)$$

$$- f_i(x_1, x_2, \cdots x_n; t)x$$

$$= \sum_{k=1}^{n} \frac{\partial}{\partial x_k} f_i(x_1, x_2, \cdots, x_n; t) \Delta x_k + \eta_i(t) \tag{10}$$

where

$$\lim_{\Delta x_{10} \to 0} \frac{\eta_i(t)}{\sqrt{\sum\limits_{k=1}^{n} \Delta x_k{}^2}} = 0 \tag{11}$$

uniformly for (X_0, t) in S. Then by (7) and (10) we may write

$$\frac{dp_i(\Delta x_{10}, t)}{dt} = \sum_{k=1}^{n} \frac{\partial}{\partial x_k} f_i(x_1, x_2, \cdots, x_n; t)p_k + \zeta_i(\Delta x_{10}, t) \tag{12}$$

where

$$\zeta_i = \frac{\eta_i}{\Delta x_{10}} \tag{13}$$

But by (9)

$$|\zeta_i| = \left| \frac{\eta_i}{\sqrt{\sum\limits_{k=1}^{n} \Delta x_k{}^2}} \right| \left| \frac{\sqrt{\sum\limits_{k=1}^{n} \Delta x_k{}^2}}{\Delta x_{10}} \right| \le \frac{e^{kb}|\eta_i|}{\sqrt{\sum\limits_{k=1}^{n} \Delta x_k{}^2}} \tag{14}$$

Hence by (11)

$$\lim_{\Delta x_{10} \to 0} |\zeta_i(t)| = 0 \tag{15}$$

uniformly for (X_0, t) in S.

Now consider the system of differential equations in n unknowns q_i, $i = 1, 2, \cdots, n$

$$\frac{dq_i}{dt} = \sum_{k=1}^{n} \frac{\partial}{\partial x_k} f_i(x_1(t), x_2(t), \cdots, x_n(t); t) q_k \qquad (16)$$

with initial conditions

$$q_i(t_0) = \begin{cases} 1, & i = 1 \\ 0, & i \neq 1 \end{cases}$$

The right-hand side of (16) is obviously continuous in the q's and is continuous in t by definition. By § 3, Lemma 1, it satisfies a Lipschitz condition on the q's. Hence (16) satisfies the hypotheses of Theorem 4. Now the system $(p_1, p_2, \cdots, p_n)$ satisfies (16) with an error

$$\epsilon = n \max |\zeta_i(t)|, \qquad i = 1, 2, \cdots, n, \qquad (X_0, t) \text{ in } S \qquad (17)$$

Hence by the proof of Theorem 5 (as in Chapter 1), $\lim_{\epsilon \to 0} p_i = q_i$; $i = 1$, $2, \cdots, n$, *uniformly* for (X_0, t) in S. Therefore by (15) and (17), $\lim p_i(t)$ exists, uniformly for (X_0, t) in S. This completes the proof. $^{\Delta x_{10} \to 0}$

From Theorem 9 we can immediately deduce the following theorem on continuity and differentiability of the solution with respect to parameters:

Theorem 10. *Consider a system of differential equations in which the functions f_i depend upon any number of parameters $\mu_1, \cdots, \mu_m$*

$$\frac{dx_i}{dt} = f_i(x_1, x_2, \cdots, x_n; \mu_1, \mu_2, \cdots, \mu_m; t), \qquad i = 1, 2, \cdots, n \qquad (18)$$

If each of the f's has partial derivatives with respect to $x_1, x_2, \cdots, x_n$; $\mu_1, \mu_2, \cdots, \mu_m$ continuous in some $(n + m + 1)$-dimensional region R, then the solutions

$$x_1(x_{10}, x_{20}, \cdots, x_{n0}; \mu_1, \mu_2, \cdots, \mu_m; t), \qquad i = 1, 2, \cdots, n \qquad (19)$$

will have partial derivatives in $\mu_1, \mu_2, \cdots, \mu_m$ continuous in all their arguments through whatever part of R the solutions (19) are defined.

Proof. Regard the parameters $\mu_1, \mu_2, \cdots, \mu_m$ as new variables, and adjoin to (18) the additional equations

$$\frac{d\mu_j}{dt} = 0, \qquad j = 1, 2, \cdots, m \qquad (20)$$

Then all the conditions of Theorem 9 are met by the system (18) + (20); hence the solutions x_i have the required properties with respect to the "initial values" of the μ's; since the μ's are taken as constants with respect to t the result follows.

We shall not need the analogues of Theorems 10 and 13 of Chapter 1. We state the two following theorems again without proof:

Theorem 11. *If in some bounded $(n + 1)$-dimensional domain R, $F(X, t)$ is continuous and locally satisfies a Lipschitz condition in X, the solution of*

$$\frac{dX}{dt} = F(X, t) \tag{21}$$

passing through any point of R may be uniquely extended arbitrarily close to the boundary of R.

Theorem 12. *If R is the domain*

$$t_1 < t < t_2, \qquad X \text{ unrestricted}$$

and if $F(X, t)$ satisfies a Lipschitz condition uniformly in every subdomain of the type

$$t_1 < t_1' \leq t \leq t_2' < t_2, \qquad all \ X$$

then the solution of (21) passing through any point of R may be extended throughout the entire open interval $t_1 < t < t_2$.

This is an immediate consequence of the fact that the solution can be extended throughout any subinterval of $t_1 < t < t_2$.

As an example, take the extremely important case of a *linear system*; i.e., a system of the form

$$\frac{dx_i}{dt} = \sum_{j=1}^{n} a_{ij}(t)x_j + b_i(t), \qquad i = 1, 2, \cdots, n \tag{22}$$

where $a_{ij}(t)$ and $b_i(t)$ are continuous in some open interval $t_1 < t < t_2$. By § 3, Lemma 1, (22) satisfies the conditions of Theorem 12; hence the values of the x's at any point t_0 (where $t_1 < t_0 < t_2$) determine a solution (of course uniquely) throughout the *whole open interval $t_1 < t < t_2$*. The study of such systems will occupy the following chapters.

Finally we remark that Picard's method may be applied to the general vector equation; the generalization of the process of Chapter 1 is perfectly straightforward. We shall see an application of this in the next chapter.

6. Systems of Higher Order

Consider the differential equation of nth order in one unknown

$$\frac{d^n x}{dt^n} = f(x^{(n-1)}, \cdots, x', x; t) \tag{1}$$

This equation may always be replaced by a system of differential equations of the first order. To do this, we take x and its first $(n-1)$ derivatives as new unknown functions $x_1, x_2, \cdots, x_n$, which must obviously satisfy the relations

$$\frac{dx_1}{dt} = x_2$$

$$\cdot \quad \cdot$$
$$\cdot \quad \cdot$$
$$\cdot \quad \cdot$$

$$\frac{dx_{n-1}}{dt} = x_n \tag{2}$$

$$\frac{dx_n}{dt} = f(x_n, \cdots, x_2, x_1; t)$$

It is easily seen that (1) and (2) are equivalent, i.e., that any solution of (2) defines a solution of (1) and vice versa. It is also clear by the existence and uniqueness theorems that wherever f satisfies a Lipschitz condition on its first n arguments, the solutions of (1) will exist and be uniquely defined by the value of x and its first $(n-1)$ derivatives at any point t_0. All our other theorems admit of obvious generalizations to the equation (1).

If in particular f is linear in the x's, i.e., if

$$f(x_n, x_{n-1}, \cdots, x_1; t) = \sum_{i=1}^{n} a_i(t)x_i + b(t) \tag{3}$$

it is clear that (2) will satisfy the conditions of Theorem 12, hence that the solutions of (1) may be extended throughout any interval in which the a's and b's are continuous.

Finally, any system of differential equations of higher order in several unknowns may be reduced to a system of the first order, provided only the system is explicitly solved for the highest derivative appearing of each unknown function

$$\frac{d^{m_i}x_i}{dt^{m_i}} = f_i(x_j^{(k)}, t) \tag{4}$$

where $k = 0, 1, \cdots, m_j - 1$; $i, j = 1, 2, \cdots, n$. It is easily seen that (4) is reducible to an equivalent first-order system of $N = \sum_{i=1}^{n} m_i$ equations, and that all our theorems likewise hold for such a system.

3

Linear Systems of Differential Equations

1. Introduction. Matrix Notation

We consider in this chapter *linear systems* of the type

$$\frac{dx_i}{dt} = \sum_{j=1}^{n} a_{ij}(t)x_j + b_i(t), \qquad i = 1, 2, \cdots, n \tag{1}$$

where $a_{ij}(t)$, $b_i(t)$ are continuous functions of t over some interval $t_1 < t < t_2$. We have already seen (Chapter 2, Theorem 12) that if t_0 is any number, $t_1 < t_0 < t_2$, then corresponding to any set of numbers $x_{10}, x_{20}, \cdots, x_{n0}$ there exists a unique solution of (1) satisfying the initial conditions

$$x_i(t_0) = x_{i0}, \qquad i = i, 2, \cdots, n \tag{2}$$

uniquely defined over the whole open interval $t_1 < t < t_2$.

If in (1) the functions $b_i(t)$ are all identically 0, the system is said to be *homogeneous*. In the homogeneous case, the system (1) is completely determined by the matrix

$$\mathscr{A}(t) = \begin{bmatrix} a_{11} & a_{12} & \cdots & a_{1n} \\ a_{21} & a_{22} & \cdots & a_{2n} \\ \cdot & \cdot & \cdots & \cdot \\ \cdot & \cdot & \cdots & \cdot \\ \cdot & \cdot & \cdots & \cdot \\ a_{n1} & a_{n2} & \cdots & a_{nn} \end{bmatrix} \tag{3}$$

where we write $\mathscr{A}(t)$ to indicate that each of the elements of the matrix (3) is a function of t. Now corresponding to any n-dimensional vector $X = (x_1, x_2, \cdots, x_n)$ and any such matrix $\mathscr{A}$ (where X and $\mathscr{A}$ may both be functions of t), we define the *linear transform of X by $\mathscr{A}$* as the vector

$$Y = (y_1, y_2, \cdots, y_n)$$

whose components y_i are given by

$$y_i = \sum_{j=1}^{n} a_{ij} x_j, \qquad i = 1, 2, \cdots, n \tag{4}$$

Symbolically we may write the relation between Y and X as

$$Y = \mathscr{A} X \tag{4'}$$

With this notation, the homogeneous case of (1) may be written as the vector equation

$$\frac{dX}{dt} = \mathscr{A}(t) X \tag{5}$$

Consider now the totality of solutions $X(t)$ of (5).

Theorem 1. *The set of solutions $X(t)$ of (5) has the properties:*

(a) $X(t) \equiv 0$ *is always a solution (the "trival solution").*

(b) *If* $X(t_0) = 0$, t_0 *some number,* $t_1 < t_0 < t_2$, *then* $X(t) \equiv 0$, $t_1 < t < t_2$. *[For $X(t) \equiv 0$ is a solution and we know that the solution is unique.]*

(c) *If $X(t)$ is a solution, then $cX(t)$ is a solution, c any constant. If $X_1(t)$ and $X_2(t)$ are solutions, then $X_1(t) + X_2(t)$ is a solution. Hence the solutions of (5) form a* linear manifold; *i.e., any linear combination of solutions is itself a solution.*

2. Linear Dependence. Fundamental Systems

We recall that m n-dimensional vectors $X_1, X_2, \cdots, X_m$ are said to be *linearly independent* if the identity

$$\sum_{i=1}^{m} c_i X_i = 0 \tag{1}$$

c_i constant, implies $c_i = 0$, $i = 1, 2, \cdots, m$. Otherwise the vectors are said to be *linearly dependent*. It is easy to see that if the vectors X_1, $X_2, \cdots, X_m$ are linearly dependent then at least one of them (say X_m) can be written as a linear combination of the others

$$X_m = \sum_{i=1}^{m-1} c_i X_i \tag{2}$$

We assume the following facts as known. No $(n + 1)$ n-dimensional vectors can be linearly independent. There exist sets of n linearly independent vectors; e.g., the *unit vectors*

$$E_1 = (1, 0, \cdots, 0), \quad E_2 = (0, 1, \cdots, 0), \cdots, \quad E_n = (0, 0, \cdots, 1) \tag{3}$$

If $X_1, X_2, \cdots, X_n$ is a linearly independent set of n vectors, then any vector can be written as a linear combination of the X's; but this is true for no set of $k < n$ vectors. A set of n vectors

$$X_i = (x_{i1}, x_{i2}, \cdots, x_{in}), \quad i = 1, 2, \cdots, n$$

is linearly independent if and only if the determinant $|x_{ij}|$ is different from 0. Any nonsingular linear transformation of a linearly independent set itself defines a linearly independent set.

A set of functions $f_i(t)$, $i = 1, 2, \cdots, m$, is said to be linearly independent *over the interval* $t_1 < t < t_2$ if the identity

$$\sum_{i=1}^{m} c_i f_i(t) \equiv 0, \quad t_1 < t < t_2 \tag{4}$$

implies $c_i = 0$, $i = 1, 2, \cdots, m$. In the case of functions whose values are *vectors*, we define linear independence in an analogous way. The c_i in this section always denote constants.

Returning now to the homogeneous linear system of § 1

$$\frac{dX}{dt} = \mathscr{A}(t)X, \quad t_1 < t < t_2 \tag{5}$$

we prove:

Theorem 2. *Let t_0 be any number, $t_1 < t_0 < t_2$, and let $X_1(t)$, $X_2(t)$, $\cdots, X_m(t)$ be any set of m solutions of (5). Then a necessary and sufficient condition that $X_i(t)$, $i = 1, 2, \cdots, m$, be linearly dependent over $t_1 < t < t_2$ is that the constant vectors $X_1(t_0), X_2(t_0), \cdots, X_m(t_0)$ be linearly dependent.*

Proof. If there exists a nontrivial relation

$$\sum_{i=1}^{m} c_i X_i(t) \equiv 0, \quad t_1 < t < t_2$$

a fortiori the same holds for $t = t_0$. Hence the necessity. To prove sufficiency, suppose there exists a nontrivial relation

$$\sum_{i=1}^{m} c_i X_i(t_0) = 0 \tag{6}$$

Consider the function

$$X(t) = \sum_{i=1}^{m} c_i X_i(t) \tag{7}$$

$X(t)$ is a solution of (5), and since it vanishes for $t = t_0$, by Theorem 1 it vanishes identically. Hence the system $X_1(t)$, $X_2(t)$, $\cdots$, $X_m(t)$ is linearly dependent.

There now follows immediately the *basic theorem on linear systems.*

Theorem 3. *No linearly independent set of solutions of* (5) *can contain more than n members. There exist linearly independent sets containing n members.*

Proof. If $X_1(t)$, $X_2(t)$, $\cdots$ $X_{n+1}(t)$ were linearly independent solutions, it would follow by Theorem 2, setting $t = t_0$, $t_1 < t_0 < t_2$, that the $(n + 1)$ constant vectors $X_i(t_0)$, $i = 1, 2, \cdots, n + 1$ would be linearly independent, which we know to be impossible. On the other hand, if we define n solutions $X_i(t)$ by

$$X_i(t_0) = E_i, \qquad i = 1, 2, \cdots, n \tag{8}$$

the E's being the unit vectors, then these X's are obviously linearly independent.

Corollary. If X_1, X_2, $\cdots$, X_n is a set of n linearly independent solutions of (5), then any solution $X(t)$ of (5) can be written in the form

$$X(t) = \sum_{i=1}^{n} c_i X_i(t) \tag{9}$$

Proof. By Theorem 3 there must exist a nontrivial linear relation among X and the X_i's

$$cX + \sum_{i=1}^{n} c_i X_i \equiv 0 \tag{10}$$

If $c = 0$, the X_i's are linearly dependent, contrary to hypothesis; hence we may divide by c and obtain the required relation.

For this reason a set of n linearly independent solutions of (5) is known as a *fundamental system* or *linear basis* of (5). The most general solution of a linear system may therefore be written as a linear combination of the members of the fundamental system. It is clear that no set of $k < n$

solutions enjoys this property. A set of n solutions of (5) forms a fundamental system if and only if the determinant of their components, $|x_{ij}|$, is different from 0. This determinant will either vanish identically in t or vanish for no value of t in the interval $t_1 < t < t_2$.

3. Solutions Expressed in Matrix Form

We shall need to introduce a few concepts of matrix algebra. If $\mathscr{A}$ and $\mathscr{B}$ are two square matrices of order n

$$\mathscr{A} = [a_{ij}], \qquad \mathscr{B} = [b_{ij}]$$

their *sum* is defined to be the matrix

$$\mathscr{A} + \mathscr{B} = [a_{ij} + b_{ij}]$$

Their *product* is the matrix $\mathscr{C}$ representing the linear transformation $Y = \mathscr{A}(\mathscr{B}X)$, i.e.

$$\mathscr{C} = \mathscr{A}\mathscr{B} = [c_{ij}], \qquad c_{ij} = \sum_{k=1}^{n} a_{ik}b_{kj}$$

Such products are known to be associative.

If the elements a_{ij} of a matrix $\mathscr{A}$ are functions of a variable t, we may define new matrices by differentiation and integration

$$\frac{d\mathscr{A}}{dt} = \left[\frac{da_{ij}}{dt}\right] \qquad \int_{t_0}^{t}\mathscr{A}\,dt = \left[\int_{t_0}^{t}a_{ij}\,dt\right]$$

It is easy to verify that

$$\frac{d(\mathscr{A} + \mathscr{B})}{dt} = \frac{d\mathscr{A}}{dt} + \frac{d\mathscr{B}}{dt}, \qquad \frac{d(\mathscr{A}\mathscr{B})}{dt} = \mathscr{A}\frac{d\mathscr{B}}{dt} + \frac{d\mathscr{A}}{dt}\mathscr{B}$$

and similar formulas. Finally we define the identity matrix $\mathscr{E}$:

$$\mathscr{E} = \begin{bmatrix} 1 & 0 & \cdots & 0 \\ 0 & 1 & \cdots & 0 \\ \cdot & \cdot & \cdots & \cdot \\ \cdot & \cdot & \cdots & \cdot \\ \cdot & \cdot & \cdots & \cdot \\ 0 & 0 & \cdots & 1 \end{bmatrix}$$

which has the property that for any vector X, $X = \mathscr{E}X$.

Suppose now that $\mathscr{A}(t)$ is an n-matrix whose components a_{ij} are continuous functions of t over some interval $t_1 < t < t_2$.

Then our homogeneous linear equation may be written

$$\frac{dX}{dt} = \mathscr{A}(t)X, \qquad X(t_0) = X_0, \qquad t_1 < t_0 < t_2 \tag{1}$$

Set

$$\mathscr{B}(t) = \int_{t_0}^{t} \mathscr{A}(t)\, dt \tag{2}$$

We consider the special case where $\mathscr{A}$ and $\mathscr{B}$ are *commutative*; i.e., that

$$\mathscr{A}\mathscr{B} \equiv \mathscr{B}\mathscr{A}, \qquad t_1 < t < t_2 \tag{3}$$

This will in particular be true if $\mathscr{A}$ is a *constant* matrix, for then

$$\mathscr{B} = (t - t_0)\mathscr{A}$$

Under this assumption, we shall obtain the solution of (1) in explicit form by Picard's approximation method.

It follows from the commutativity of $\mathscr{A}$ and $\mathscr{B}$ that

$$\frac{d\mathscr{B}^m}{dt} = m\mathscr{A}\mathscr{B}^{m-1}, \qquad m = 1, 2, \cdots \tag{4}$$

Let the initial value X_0 be the first approximation. Then the next approximation, $X_1(t)$, is defined by

$$\frac{dX_1}{dt} = \mathscr{A}X_0, \qquad X_1(t_0) = X_0 \tag{5}$$

i.e., by

$$X_1(t) = X_0 + \int_{t_0}^{t} \mathscr{A}X_0\, dt = (\mathscr{E} + \mathscr{B})X_0 \tag{6}$$

The mth approximating solution, $X_m(t)$, will satisfy

$$\frac{dX_m}{dt} = \mathscr{A}X_{m-1}, \qquad X_m(t_0) = X_0 \tag{7}$$

Using (4) and (6), it follows easily by induction that

$$X_m(t) = \mathscr{E} + \mathscr{B} + \frac{\mathscr{B}^2}{2!} + \cdots + \frac{\mathscr{B}^m}{m!}\, X_0 \tag{8}$$

But by the proof of Picard's theorem (as generalized from the special case of Chapter 1) it follows that as $m \to \infty$, $X_m(t)$ converges uniformly in a sufficiently small neighborhood of t_0 to a solution $X(t)$ of (1).

If we set $X_0 = E_i$, the ith unit vector, the right-hand side of (8) reduces to the ith column of the matrix

$$\mathscr{E} + \sum_{k=1}^{m} \frac{\mathscr{B}^k}{k!} \tag{9}$$

whence it appears that *each of the terms* of (9) approaches a limit *uniformly in t* as $m \to \infty$. The matrix so defined we call for obvious symbolic reasons $e^{\mathscr{B}}$. Then the solution of (1) may be displayed in the form

$$X(t) = e^{\mathscr{B}} X_0 = \exp \left\{ \int_{t_0}^{t} \mathscr{A} \, dt \right\} X_0 \tag{10}$$

In particular, the n columns of $e^{\mathscr{B}}$ considered as vectors constitute a fundamental system of (1).

Expressing the condition that (10) be a solution of (1), we have

$$\frac{d(e^{\mathscr{B}})}{dt} X_0 = \mathscr{A} e^{\mathscr{B}} X_0 \tag{11}$$

and substituting $X_0 = \mathscr{E}$, (11) reduces to

$$\frac{d(e^{\mathscr{B}})}{dt} = \mathscr{A} e^{\mathscr{B}} \tag{12}$$

This is of course in agreement with our symbolic notation, since $d\mathscr{B}/dt = \mathscr{A}$. Observe the analogy between (10) and the solution of the simple differential equation

$$\frac{dx}{dt} = a(t)x, \qquad x(t_0) = x_0$$

namely

$$x = x_0 \exp \left\{ \int_{t_0}^{t} a(t) \, dt \right\} \tag{13}$$

This is of course the special case of (10) when $n = 1$.

It should be remarked that the series

$$\mathscr{E} + \sum_{m=1}^{\infty} \frac{\mathscr{B}^m}{m!} \tag{14}$$

defining $e^{\mathscr{B}}$ not only converges in the restricted interval of Picard's theorem, but converges uniformly over any closed interval in which each of the elements of $\mathscr{A}(t)$ is continuous. In fact, if h is the length of the interval, and if in the interval each of the elements of $\mathscr{A}(t)$ is bounded by some constant M, the reader will prove without difficulty that each of the elements of $\mathscr{B}^m$ is bounded by $n^{m-1}(Mh)^m$. Hence it will follow that over the given interval (14) converges uniformly. Since the differentiated series has the same property, formula (12) is still valid. Hence (10) is easily seen to give the solution of (1) over the entire interval in which $\mathscr{A}(t)$ is continuous, i.e., in which the solution of (1) exists.

It should be emphasized that this section is based on the assumption that (3) holds.

4. Reduction of Order of a System

Suppose $k < n$ linearly independent solutions of the homogeneous system

$$\frac{dx_i}{dt} = \sum_{j=1}^{n} a_{ij}(t)x_j, \qquad i = 1, 2, \cdots, n \tag{1}$$

have somehow been obtained. We shall show that (1) may be "reduced to a system of order $n - k$" in the following sense; if a fundamental system of solutions of a certain system of order $n - k$ is known, a fundamental system of (1) may be determined by quadratures alone.

Theorem 4. *Suppose there has been found one nontrivial solution* $X(t)$ *of* (1)

$$X(t) = \{\eta_1(t), \eta_2(t), \cdots, \eta_n(t)\} \tag{2}$$

Assume one of its components (say without loss of generality η_n) *vanishes nowhere on* $t_1 < t < t_2$.* *Then* (1) *may be reduced to a system of order* $n - 1$.

Proof. Define new variables y_i, $i = 1, 2, \cdots, n$, by

$$y_i = x_i - \frac{\eta_i(t)}{\eta_n(t)}x_n, \qquad i = 1, 2, \cdots, n - 1, \qquad y_n = \frac{1}{\eta_n(t)}x_n \tag{3}$$

This is possible since $\eta_n(t)$ vanishes nowhere, and the transformation is nonsingular since we can write

$$x_i = y_i + \eta_i(t)y_n, \qquad i = 1, 2, \cdots, n - 1; \qquad x_n = \eta_n(t)y_n \tag{4}$$

We wish to express the y's as solutions of a homogeneous system. Differentiating (3) we obtain

$$\frac{dy_i}{dt} = \frac{dx_i}{dt} - x_n \frac{d}{dt}\left(\frac{\eta_i}{\eta_n}\right) - \frac{\eta_i}{\eta_n}\frac{dx_n}{dt}, \qquad i = 1, 2, \cdots, n - 1$$

$$\frac{dy_n}{dt} = \frac{1}{\eta_n}\frac{dx_n}{dt} + \frac{d}{dt}\left(\frac{1}{\eta_n}\right)x_n \tag{5}$$

Substituting in (5) first for dx_i/dt from (1) and then for x_i from (4), the y's are seen to satisfy a system of the form

$$\frac{dy_i}{dt} = \sum_{j=1}^{n} b_{ij}(t)y_j, \qquad i = 1, 2, \cdots, n \tag{6}$$

where the b's may be explicitly calculated; they are certainly continuous over $t_1 < t < t_2$. It is easy to see that (1) and (6) are equivalent; i.e.,

* This will in any case be true over some sufficiently small interval.

that to every set of x's satisfying (1) there is defined by (3) a set of y's satisfying (6), and vice versa. The solution of (6) corresponding to (2) is

$$y_i = \begin{cases} 0, & i \neq n \\ 1, & i = n \end{cases} \tag{7}$$

Substituting (7) in (6) we obtain, without making a long calculation, the relation

$$b_{in}(t) \equiv 0, \qquad i = 1, 2, \cdots, n \tag{}$$

Therefore in particular the first $(n-1)$ equations of (6) do not contain y_n at all; hence they may be treated as a separate linear system of $(n-1)$ equations in the $(n-1)$ unknowns $y_1, y_2, \cdots, y_{n-1}$. Suppose a fundamental system

$$Z_i = (y_{i1}, y_{i2}, \cdots, y_{i,n-1}), \qquad i = 1, 2, \cdots, n-1 \tag{8}$$

of this set of equations is known. Then functions y_{in}, $i = 1, 2, \cdots$, $n - 1$, may be defined by quadratures from

$$\frac{dy_{in}}{dt} = \sum_{j=1}^{n-1} b_{nj}(t) y_{ij}(t) \tag{9}$$

If we adjoin the functions y_{in} as nth components to the vectors Z_i, we construct a set of $(n-1)$ n-dimensional vectors $Y_1, Y_2, \cdots, Y_{n-1}$. They are by construction solutions of (6), and a fortiori linearly independent. Let us now adjoin to this set the vector $Y_n = (0, 0, \cdots, 1)$. Then $Y_1, Y_2, \cdots, Y_n$ are all solutions of (6). Suppose we could write

$$\sum_{i=1}^{n} c_i Y_i \equiv 0$$

The first $(n-1)$ c's must be 0, since the Z's are linearly independent. Since $Y_n \neq 0$, it follows that $c_n = 0$, and the Y's are linearly independent. If finally we define from the Y's a system of n vectors $X_1, X_2, \cdots, X_n$ by (4), these will all be solutions of (1), and the proof is complete.

Suppose now there is given another solution

$$X(t) = (\zeta_1, \zeta_2, \cdots, \zeta_n) \tag{10}$$

of (1) linearly independent of $(\eta_1, \eta_2, \cdots, \eta_n)$. Then a particular solution of the first $(n-1)$ equations of (6) is

$$Z = (\theta_1, \theta_2, \cdots, \theta_{n-1})$$

where

$$\theta_i = \zeta_i - \frac{\eta_i}{\eta_n} \zeta_n, \qquad i = 1, 2, \cdots, n-1 \tag{11}$$

If $\theta_i \equiv 0$, solutions (2) and (10) are linearly dependent; hence on some interval one of the θ's never vanishes. Then we may apply our method again on (6) using the particular solution (11) and reduce the order of (1) to $n - 2$.

Proceeding in this way, it is easy to prove that, given $k < n$ linearly independent solutions of (1), there is some interval in which (1) may be reduced to a system of order $n - k$. In fact, any closed interval of $t_1 < t < t_2$ may be covered by a finite set of such intervals, and the fundamental systems defined in each covering interval may be "matched" so as to define a fundamental system over the whole closed interval. Thus the problem of reduction of order is substantially solved.

5. Nonhomogeneous Systems

Consider now the nonhomogeneous linear system

$$\frac{dx_i}{dt} = \sum_{j=1}^{n} a_{ij}(t)x_j + b_i(t), \qquad i = 1, 2, \cdots, n \tag{1}$$

or in vector notation

$$\frac{dX}{dt} = \mathscr{A}X + B, \qquad B = (b_1, b_2, \cdots, b_n) \tag{1'}$$

Let $a_{ij}(t)$, $b_i(t)$ be continuous on $t_1 < t < t_2$. Suppose $X_1(t)$ is any particular solution of (1). Then if $X_2(t)$ is any other solution of (1), obviously $X(t) = X_2(t) - X_1(t)$ is a solution of the homogeneous equation

$$\frac{dx_i}{dt} = \sum_{j=1}^{n} a_{ij}(t)x_j$$

i.e., of

$$\frac{dX}{dt} = \mathscr{A}X \tag{2}$$

Theorem 5. *The most general solution of the nonhomogeneous equation* (1) *is obtained by adding to any particular solution of* (1) *the general solution of the homogeneous equation* (2).

Therefore if the homogeneous equation has been solved, all we need do to find the general solution of (1) is find one particular solution of (1).

Theorem 6. *A particular solution of the nonhomogeneous system* (1) *may be obtained from a fundamental system of* (2) *by means of quadratures alone.*

Proof. Let $X_1, X_2, \cdots, X_n$ be a fundamental system of (2). We use the so-called "method of variation of constants." Suppose a particular solution of (1) can be written in the form

$$X = \sum_{i=1}^{n} c_i X_i \qquad (3)$$

where the c's are *not constants* as in the homogeneous case, but *unknown functions of t*. We shall try to satisfy (1) by a suitable choice of the functions c_i. Differentiating (3) and substituting in (1) we have

$$\sum_{i=1}^{n} c_i \frac{dX_i}{dt} + \sum_{i=1}^{n} X_i \frac{dc_i}{dt} = \sum_{i=1}^{n} c_i \mathscr{A} X_i + B \qquad (4)$$

or since the X_i's satisfy (2)

$$\sum_{i=1}^{n} X_i \frac{dc_i}{dt} = B \qquad (5)$$

If we now set

$$X_i = (x_{i1}, x_{i2}, \cdots, x_{in}) \qquad (6)$$

(5) may be written in the form

$$\sum_{j=1}^{n} x_{ji} \frac{dc_j}{dt} = b_i, \qquad i = 1, 2, \cdots, n \qquad (7)$$

Since $|x_{ij}|$ is nowhere 0 (end of § 2), (7) may everywhere be solved for dc_i/dt. Thus

$$\frac{dc_i}{dt} = p_i(t) \qquad (8)$$

where $p_i(t)$ are continuous functions that may be easily calculated. Equation (8) may be solved by quadratures; and any solution $c_1, c_2, \cdots, c_n$ of (8) will define a particular solution of our nonhomogeneous system (1) by means of (4). This completes the proof.

PART B.
LINEAR EQUATIONS OF HIGHER ORDER

6. Fundamental Systems

We consider now the linear differential equation of order n in one unknown

$$\frac{d^n x}{dt^n} + p_1(t) \frac{d^{n-1} x}{dt^{n-1}} + \cdots + p_n(t)x + q(t) = 0 \qquad (1)$$

where the functions $p_i(t)$ and $q(t)$ are continuous over some interval $t_1 < t < t_2$. We are interested only in *normalized* equations where the coefficient of $d^n x/dt^n$ is 1. If $d^n x/dt^n$ has a coefficient $p_0(t)$ which vanishes nowhere, the equation may be normalized by dividing by p_0.

We consider at first the homogeneous case in which $q(t) \equiv 0$. We shall generally use the notation

$$L(x) = \frac{d^n x}{dt^n} + \sum_{i=1}^{n} p_i(t) \frac{d^{n-1}x}{dt^{n-1}} = 0 \tag{2}$$

Following the method of Chapter 2, § 6, we may replace (2) by the linear system

$$\frac{dx_1}{dt} = x_2$$

$$\frac{dx_2}{dt} = x_3$$

$$\tag{3}$$

$$\frac{dx_{n-1}}{dt} = x_n$$

$$\frac{dx_n}{dt} = -p_1 x_n - p_2 x_{n-1} - \cdots - p_n x_1$$

or in vector form

$$\frac{dX}{dt} = \mathscr{A} X \tag{3'}$$

where the matrix $\mathscr{A}$ is

$$\mathscr{A}(t) = \begin{bmatrix} 0 & 1 & 0 & \cdots & 0 & 0 \\ 0 & 0 & 1 & \cdots & 0 & 0 \\ \cdot & \cdot & \cdot & \cdots & \cdot & \cdot \\ \cdot & \cdot & \cdot & \cdots & \cdot & \cdot \\ \cdot & \cdot & \cdot & \cdots & \cdot & \cdot \\ \cdot & \cdot & \cdot & \cdots & \cdot & \cdot \\ 0 & 0 & 0 & \cdots & 0 & 1 \\ -p_n & -p_{n-1} & -p_{n-2} & \cdots & -p_2 & -p_1 \end{bmatrix} \tag{4}$$

Then as we have seen in Chapter 2, § 6, (2) and (3) are equivalent. For if $x = \phi(t)$ is any solution of (2), then

$$X(t) = (\phi, \phi', \cdots \phi^{[n-1]}) \tag{5}$$

is a solution of (3'); and conversely, if $X(t)$ is a solution of (3'), its first component is a solution of (2). It follows from the remarks in Chapter 2, § 6, and the results of Part A of this chapter, that the values of x and its first $(n - 1)$ derivatives at some point t_0 (where $t_1 < t_0 < t_2$) define a unique solution of (2) over the whole open interval $t_1 < t < t_2$. If at any point $t = t_0$, x and its first $(n - 1)$ derivatives all vanish, $x(t)$ is identically 0. The solutions $x(t)$ of (2) form a linear manifold.

It is easy to see that the correspondence we have set up between the solutions of (2) and (3) *preserves linear independence.* For if $\phi_1(t)$, $\phi_2(t)$, $\cdots$, $\phi_m(t)$ is a linearly independent set of solutions of (2), a fortiori the m vectors

$$X_i(t) = (\phi_i, \phi_i', \cdots \phi_i^{[n-1]}), \qquad i = 1, 2, \cdots, m \tag{6}$$

are linearly independent. On the other hand, if there exists a nontrivial linear relation

$$\sum_{i=1}^{m} c_i \phi_i(t) \equiv 0$$

differentiating we obtain

$$\sum_{i=1}^{m} c_i \phi_i^{[j-1]}(t) \equiv 0, \qquad j = 1, 2, \cdots, n \tag{7}$$

and the equations (7) are simply the components of

$$\sum_{i=1}^{m} c_i X_i \equiv 0 \tag{8}$$

Therefore the X's also are linearly dependent. This means that the ϕ's are linearly independent if and only if the X's are.

Hence as in § 2 it follows that there can exist no set of $(n + 1)$ linearly independent solutions of (2), but that there exist sets of n linearly independent solutions of (2). Any such set will be a fundamental system of (2). Take a point t_0, $t_1 < t_0 < t_2$. If $\phi_1(t)$, $\phi_2(t)$, $\cdots$, $\phi_n(t)$ are any n solutions of (2), they constitute a fundamental system if and only if the determinant of the components of the X's taken at $t = t_0$ fails to vanish

$$W(\phi_1, \phi_2, \cdots, \phi_n) = \begin{vmatrix} \phi_1 & \phi_2 & \cdots & \phi_n \\ \phi_1' & \phi_2' & \cdots & \phi_n' \\ \cdot & \cdot & \cdots & \cdot \\ \cdot & \cdot & \cdots & \cdot \\ \cdot & \cdot & \cdots & \cdot \\ \phi_1^{[n-1]} & \phi_2^{[n-1]} & \cdots & \phi_n^{[n-1]} \end{vmatrix} \neq 0 \tag{9}$$

This determinant is known as the *Wronsky determinant* or *Wronskian* of the functions $\phi_1, \phi_2, \cdots, \phi_n$. We may then state our results thus:

Theorem 7. The homogeneous equation (2) *always has a fundamental system of precisely n solutions. A set of n solutions* $\phi_1(t), \phi_2(t), \cdots, \phi_n(t)$ *of* (2) *constitutes a fundamental system if and only if its Wronskian* $W(\phi_1, \phi_2, \cdots, \phi_n)$ *fails to vanish for some point* t_0, $t_1 < t_0 < t_2$. *The Wronskian will either be different from* 0 *everywhere on this interval or vanish identically.*

7. The Wronsky Determinant

We shall now consider the significance of the Wronksian of any *n arbitrary* functions (not necessarily solutions of some differential equation). It will appear that the vanishing of the Wronskian is related to their linear dependence; but the criterion is not as simple as in the special case of Theorem 7.

Theorem 8. Let $\phi_1(t), \phi_2(t), \cdots, \phi_n(t)$ *be any n functions continuous together with their first* $(n - 1)$ *derivatives over some interval* $t_1 < t < t_2$.

(a) *If the ϕ's are linearly dependent over* $t_1 < t < t_2$, *then*

$$W(\phi_1, \phi_2, \cdots, \phi_n) \equiv 0, \qquad t_1 < t < t_2$$

(b) *Suppose*

(1) $W(\phi_1, \phi_2, \cdots, \phi_n) \equiv 0, t_1 < t < t_2;$ *but*

(2) *for some* $(n - 1)$ *of the ϕ's (say without loss of generality, all but ϕ_n)*

$$W(\phi_1, \phi_2, \cdots, \phi_{n-1}) \neq 0, \text{ all } t, \qquad t_1 < t < t_2 \tag{1}$$

then $\phi_1, \phi_2, \cdots, \phi_n$ *are linearly dependent over* $t_1 < t < t_2$.

Proof.

(a) If there exists a nontrivial relation

$$\sum_{i=1}^{n} c_i \phi_i(t) \equiv 0$$

Differentiating we obtain

$$\sum_{i=1}^{n} c_i \phi_i^{[j-1]}(t) \equiv 0, \qquad j = 1, 2, \cdots, n \tag{2}$$

Equation (2) implies that for $t = t_0$, each t_0, the determinant of the c's will vanish. This determinant is precisely $W(\phi_1, \phi_2, \cdots, \phi_n)$.

(b) The equation

$$W(\phi_1, \phi_2, \cdots, \phi_{n-1}, x) \equiv 0 \tag{3}$$

is a linear homogeneous differential equation of order $(n - 1)$ in x with continuous coefficients. The coefficient of $(d^{n-1}x)/(dt^{n-1})$ is $W(\phi_1, \phi_2, \cdots, \phi_{n-1})$, and since by (1) this never vanishes, (3) may be normalized.

Now $\phi_1, \phi_2, \cdots, \phi_{n-1}$ are all solutions of (3) since they make two columns of the determinant equal. $x = \phi_n$ is a solution by hypothesis. Since (3) is only of order $(n - 1)$, its n solutions $\phi_1, \phi_2, \cdots, \phi_n$ must be linearly dependent.

Part (b) of Theorem 8 may be proved directly without the use of the theory of differential equations, but the proof will be longer. The second part of the hypothesis in (b) cannot be omitted, but we may prove the corollary:

Corollary. Let $\phi_1(t), \phi_2(t), \cdots \phi_n(t)$ be any n functions continuous together with their first $(n - 1)$ derivatives over some interval $t_1 < t < t_2$. Let

$$W(\phi_1, \phi_2, \cdots, \phi_n) \equiv 0, \qquad t_1 < t < t_2 \tag{4}$$

Then there exists a subinterval over which $\phi_1, \phi_2, \cdots, \phi_n$ are linearly dependent.

Proof. We give a proof by induction. The theorem is trivial for $n = 1$. Suppose it true for $n = m - 1$. Then consider the m functions $\phi_1, \phi_2, \cdots, \phi_m$. If the Wronskian $W(\phi_1, \phi_2, \cdots \phi_{m-1})$ fails to vanish for some point $t_0, t_1 < t_0 < t_2$, there is some interval of t_0 in which it does not vanish, and Theorem 8 applies. Otherwise the smaller Wronskian is identically 0, and by our induction $\phi_1, \phi_2, \cdots, \phi_{m-1}$ are linearly dependent over some subinterval of $t_1 < t < t_2$. This completes the proof.

8. Further Properties of the Fundamental System

We have seen that to any normalized linear homogeneous equation of nth order there corresponds a fundamental system of n solutions whose Wronskian never vanishes. We can prove the following converse.

Theorem 9. *Let $\phi_1(t), \phi_2(t), \cdots, \phi_n(t)$ be any set of functions continuous together with their first n derivatives over some interval $t_1 < t < t_2$; and let their Wronskian vanish nowhere. Then there exists a unique normalized equation of nth order*

$$L(x) = 0, \qquad t_1 < t < t_2 \tag{1}$$

possessing $\phi_1, \phi_2, \cdots, \phi_n$ as a fundamental system.

Proof. The equation

$$L(x) = \frac{W(\phi_1, \phi_2, \cdots, \phi_n, x)}{W(\phi_1, \phi_2, \cdots, \phi_n)} = 0 \tag{2}$$

is a homogeneous linear equation of nth order in x with continuous coefficients. The coefficient of $(d^n x)/(dt^n)$ is 1. Equation (2) has the n

solutions $\phi_1, \phi_2, \cdots, \phi_n$; and by Theorem 8 the ϕ's are linearly independent, since their Wronskian does not vanish. Therefore (2) satisfies all the conditions and the existence is proved.

Suppose now there exist two such equations

$$L_1(x) = 0, \qquad L_2(x) = 0$$

Consider the equation

$$M(x) = L_1(x) - L_2(x) = 0 \tag{3}$$

It is a homogeneous linear equation with continuous coefficients. If any one of its coefficients is anywhere different from 0, there exists some interval over which it can be normalized. But its order is at most $(n - 1)$, and it possesses the n solutions $\phi_1, \phi_2, \cdots, \phi_n$ which are linearly independent over any interval (their Wronskian being nowhere 0). Therefore $M(x)$ cannot be normalized; i.e., it vanishes identically. Hence the equation (1) is unique.

Theorem 10. *Let* $\phi_1, \phi_2, \cdots, \phi_n$ *be any fundamental system of*

$$L(x) = \frac{d^n x}{dt^n} + \sum_{i=1}^{n} p_i(t) \frac{d^{n-i} x}{dt^{n-i}} = 0, \qquad t_1 < t < t_2 \tag{4}$$

Set $W(t) = W(\phi_1, \phi_2, \cdots, \phi_n)$. *Then if* t_0 *is any number,* $t_1 < t_0 < t_2$

$$W(t) = W(t_0) \exp\left\{ -\int_{t_0}^{t} p_1(t)\, dt \right\} \tag{5}$$

Thus the value of the Wronskian is essentially independent of the particular fundamental system we choose. This result explains the fact that in Theorem 7 we have only to assume that $W(t)$ fails to vanish at *one* point.

Proof. By virtue of the uniqueness of $L(x)$ in Theorem 9 we may write

$$L(x) \equiv \frac{W(\phi_1, \phi_2, \cdots, \phi_n, x)}{W(\phi_1, \phi_2, \cdots, \phi_n)} \tag{6}$$

The identity in (6) implies that the coefficients of each term on either side are identical. In particular, equating the coefficients of $(d^{n-1}x)/(dt^{n-1})$, we have

$$p_1(t) = -\frac{1}{W(t)} \begin{vmatrix} \phi_1 & \phi_2 & \cdots & \phi_n \\ \phi_1' & \phi_2' & \cdots & \phi_n' \\ \cdot & \cdot & \cdots & \cdot \\ \cdot & \cdot & \cdots & \cdot \\ \phi_1^{[n-2]} & \phi_2^{[n-2]} & \cdots & \phi_n^{[n-2]} \\ \phi_1^{[n]} & \phi_2^{[n]} & \cdots & \phi_n^{[n]} \end{vmatrix} \tag{7}$$

Now the determinant in (7) is exactly $[dW(t)]/dt$; for by the well-known formula for the derivative of a determinant

$$W'(t) = \begin{vmatrix} \phi_1' & \phi_2' & \cdots & \phi_n' \\ \phi_1' & \phi_2' & \cdots & \phi_n' \\ \cdot & \cdot & \cdots & \cdot \\ \cdot & \cdot & \cdots & \cdot \\ \cdot & \cdot & \cdots & \cdot \\ \phi_1^{[n-1]} & \phi_2^{[n-1]} & \cdots & \phi_n^{[n-1]} \end{vmatrix} + (n-2) \text{ terms}$$

$$+ \begin{vmatrix} \phi_1 & \phi_2 & \cdots & \phi_n \\ \cdot & \cdot & \cdots & \cdot \\ \cdot & \cdot & \cdots & \cdot \\ \cdot & \cdot & \cdots & \cdot \\ \phi_1^{[n-2]} & \phi_2^{[n-2]} & \cdots & \phi_n^{[n-2]} \\ \phi_1^{[n]} & \phi_2^{[n]} & \cdots & \phi_n^{[n]} \end{vmatrix} \tag{8}$$

where the first $(n-1)$ terms vanish identically, having two equal rows. Therefore

$$p_1(t) = -\frac{W'(t)}{W(t)} \tag{9}$$

and integrating (9) we obtain (5). This completes the proof.

9. Reduction of Order

We recapitulate briefly the theory of § 4 in the case of a single equation of higher order.

Theorem 11. *Let $\eta(t)$ be a nontrivial solution of*

$$L(x) = \frac{d^n x}{dt^n} + \sum_{i=1}^{n} p_i(t) \frac{d^{n-i} x}{dt^{n-i}} \tag{1}$$

Then the transformation

$$x = \eta y \tag{2}$$

reduces (1) to an equation of order $(n-1)$ in $z = y'$ over any interval in which η vanishes nowhere.

Proof. Differentiating (2) and substituting in (1), we obtain

$$L(\eta y) = y^{[n]} \eta + \text{lower order terms}$$

Dividing by η we see that y satisfies an equation of the form

$$y^{[n]} + \sum_{i=1}^{n} q_i(t) y^{[n-i]} = 0 \tag{3}$$

where the q's are continuous. Equations (1) and (3) are equivalent, and since $y \equiv 1$ is a solution of (3), $q_n \equiv 0$. Hence if we set $z = y'$, we obtain an equation of order $(n-1)$ in z

$$z^{[n-1]} + q_1 z^{[n-2]} + \cdots + q_{n-1} z = 0 \tag{4}$$

Suppose a fundamental system of (4) has been obtained. We may then define $(n-1)$ solutions of (3) from this fundamental system, and adjoin to these the solution $y \equiv 1$. The proof that the n solutions of (3) so defined are linearly independent follows easily from $z = y'$; the reader can verify the fact directly. By (2) we may then define n solutions of (1), which will be linearly independent [(2) being nonsingular] and hence constitute a fundamental system of (1). This completes the proof.

As before, it may easily be shown that, given $k < n$ linearly independent solutions of (1), (1) may be reduced to a system of order $(n-k)$.

10. The Nonhomogeneous Case

We consider finally the nonhomogeneous equation of nth order

$$L(x) + q(t) = 0, \qquad t_1 < t < t_2 \tag{1}$$

$$L(x) = \frac{d^n x}{dt^n} + \sum_{i=1}^{n} p_i(t) x^{[n-i]}$$

Equation (1) is equivalent to the system

$$\frac{dx_1}{dt} = x_2$$

$$\begin{matrix} \cdot & \cdot \\ \cdot & \cdot \\ \cdot & \cdot \end{matrix} \tag{2}$$

$$\frac{dx_{n-1}}{dt} = x_n$$

$$\frac{dx_n}{dt} = -p_n x_1 - \cdots - p_1 x_n - q(t)$$

Exactly as in Theorem 5, we prove that the most general solution of (1) may be obtained by adding to a particular solution of (1) the general solution of $L(x) = 0$. Referring to Theorem 6 and making the appropriate substitutions, we can apply the method of "variation of constants" to obtain a particular solution of (1).

Theorem 12. Let $\phi_1, \phi_2, \cdots, \phi_n$ be a fundamental system of $L(x) = 0$. Then a particular solution $\psi(t)$ of the nonhomogeneous equation is given by

$$\psi(t) = \sum_{i=1}^{n} c_i(t)\phi_i(t) \tag{3}$$

where the c's satisfy the system

$$\sum_{i=1}^{n} c_i'(t)\phi_i^{[j-1]}(t) = 0, \qquad j = 1, 2, \cdots, n-1$$

$$\sum_{i=1}^{n} c_i'(t)\phi_i^{[n-1]}(t) = -q(t) \tag{4}$$

The proof follows directly from Theorem 6; however, the following direct verification is interesting. The determinant of the coefficients of c_i' in (4) is precisely $W(\phi_1, \phi_2, \cdots, \phi_n)$. Since by hypothesis this vanishes nowhere, (4) may everywhere be solved, and the c's determined by quadratures. From (3) and (4) it appears immediately that

$$\psi^{[k]}(t) = \sum_{i=1}^{n} c_i(t)\phi_i^{[k]}(t), \qquad k = 0, 1, \cdots, n-1$$

$$\psi^{[n]}(t) = \sum_{i=1}^{n} c_i(t)\phi_i^{[n]}(t) - q(t) \tag{5}$$

Setting $x = \psi$ in (1) and using (5), it follows that

$$L(\psi) = \sum_{i=1}^{n} c_i(t)L(\phi_i) - q(t)$$

Since $L(\phi_i) = 0$, we obtain

$$L(\psi) = -q(t), \qquad t_1 < t < t_2$$

which completes the proof.

11. Green's Function

We shall consider in this section a method whereby the solution of the nonhomogeneous linear equation

$$L(x) + q(t) = 0 \tag{1}$$

where

$$L(x) = \frac{d^n x}{dt^n} + \sum_{i=1}^{n} p_i(t)x^{[n-i]}$$

may be obtained from a single function associated with $L(x)$ known as *Green's function* for $L(x)$. It will be more convenient for this problem to assume that $p_i(t)$ and $q(t)$ are continuous over some *closed* interval $a \leq t \leq b$.

A solution $x(t)$ of $L(x) = 0$ satisfying the initial conditions

$$x(a) = 0, \qquad x'(a) = 0, \cdots, x^{[n-1]}(a) = 0$$

must vanish identically, provided we assume (as we have up to now) that $x(t)$ and its first $(n-1)$ derivatives are *continuous* over $[a, b]$. Let us drop this last assumption and consider a solution $x(t)$ whose $(n-1)$st derivative $x^{[n-1]}(t)$ has a discontinuity at some point s, $a \leq s \leq b$. Precisely we consider the function given by the following definition:

Definition 1. *Let $x(t)$ be a function defined over $a \leq t \leq b$ such that*

(1) *$x(t)$ and its first n derivatives exist and satisfy $L(x) = 0$ for every point of $a \leq t \leq b$ except possibly the single point $t = s$.*

(2) $x(a) = 0,\ x'(a) = 0, \cdots, x^{[n-1]}(a) = 0.$ (2)

(3) *The functions $x(t)$, $x'(t), \cdots, x^{[n-2]}(t)$ are continuous at s (hence everywhere on $a \leq t \leq b$). $x^{[n-1]}(t)$ has a discontinuity at $t = s$*

$$x^{[n-1]}(s - 0) = 0, \qquad x^{[n-1]}(s + 0) = -1 \qquad (3)$$

Then $x(t)$ is called a Green's function or weighting function for $L(x)$ and is denoted by $K(s, t)$.

Let L_t denote differentiation with respect to t by the operator L. It is clear that $K(s, t)$ exists and is unique; for over $a \leq t \leq s$, $K(s, t) \equiv 0$; and over $s \leq t \leq b$, K and its first n derivatives are continuous and satisfy $L_t(K) = 0$, whence K is uniquely determined by the boundary conditions (2) and (3).

Lemma 1. *$K(s, t)$ and its first $(n-2)$ derivatives with respect to t are continuous in s and t simultaneously, for $a \leq s \leq b$, $a \leq t \leq b$. The same is true for $K_t^{[n-1]}(s, t)$ and $K_t^{[n]}(s, t)$, except that these functions may be discontinuous where $s = t$.*

Proof. For $s > t$, $K(s, t) \equiv 0$ and the lemma is trivial. For $s \leq t$, set

$$x_i(s, t) = K_t^{[i-1]}(s, t), \qquad i = 1, 2, \cdots, n$$

Then by a change of variable the x's are seen to satisfy the system

$$\frac{dx_1}{dt} = x_2$$

$$\cdot \qquad \cdot$$

$$\cdot \qquad \cdot$$

$$\cdot \qquad \cdot \qquad (4)$$

$$\frac{dx_{n-1}}{dt} = x_n$$

$$\frac{dx_n}{dt} = -p_n(t + s - a)x_1 - \cdots - p_1(t + s - a)x_n$$

for $a \leq t \leq b + a - s$, with the initial conditions

$$x_i(a) = 0, \qquad i = 1, 2, \cdots, n - 1; \qquad x_n(a) = -1$$

Set

$$X(s, t) = \{x_1(s, t), \cdots, x_n(s, t)\}$$

Let $s_1 < s_2$ be any two values of s, and set $s = s_1$ in (4). Then $X(s_2, t)$ satisfies (4) with a certain error ϵ; and since the p's are continuous and the x's all bounded, it is easy to see that $\lim_{s_2 \to s_1} \epsilon = 0$, uniformly in t. Hence by the fundamental convergence theorem, $\lim_{s_2 \to s_1} X(s_2, t) = X(s_1, t)$, uniformly in t. That is to say, $X(s, t)$ is continuous in s and t simultaneously, $s \leq t$. Since $K(s, t)$ and its first $(n - 2)$ derivatives are continuous also at $s = t$, they are continuous everywhere in $a \leq s \leq b$, $a \leq t \leq b$, and Lemma 1 follows.

We now can prove our main result;

Theorem 13. *The solution $x(t)$ of the nonhomogeneous equation* (1) *with initial conditions*

$$x^{[i-1]}(a) = 0, \qquad i = 1, 2, \cdots, n \tag{5}$$

is given by

$$x(t) = \int_a^b K(s, t) q(s)\, ds \tag{6}$$

Proof. By Lemma 1 the integral (6) exists. Since the derivatives of $K(s, t)$ with respect to t up to the $(n - 2)$nd order are continuous in s, we may differentiate under the integral sign $(n - 1)$ times with respect to t, obtaining

$$x^{[j-1]}(t) = \int_a^b K_t^{[j-1]}(s, t) q(s)\, ds, \qquad j = 1, 2, \cdots, n \tag{7}$$

The fact that $K_t^{[n-1]}(s, t)$ has a single discontinuity at $s = t$ does not of course impair the validity of the last step. Since $K_t^{[n-1]}(s, t) \equiv 0$, $s > t$, we may also write

$$x^{[n-1]}(t) = \int_a^t K^{[n-1]}(s, t) q(s)\, ds \tag{8}$$

Since the integrand is now continuous over the whole range of integration, we may differentiate again by the usual rule, obtaining

$$x^{[n]}(t) = \int_a^t K^{[n]}(s, t) q(s)\, ds + K^{[n-1]}(s, t) q(s)]_{s=t-0} \tag{9}$$

Since

$$K^{[n-1]}(s, t)]_{s=t-0} = K^{[n-1]}(s, t)]_{t=s+0} = -1$$

by (3), we may write (9) in the form

$$x^{[n]}(t) = \int_a^b K^{[n]}(s, t)q(s)\, ds - q(t) \tag{10}$$

From (7) and (10) we obtain

$$L(x) = \int_a^b L_t(K)q(s)\, ds - q(t) = -q(t), \qquad a \le t \le b \tag{11}$$

Hence $x(t)$ everywhere satisfies the nonhomogeneous equation (1). From (7) the boundary conditions are obviously satisfied, and $x(t)$ is the required function.

We can give a physical interpretation of Green's function in the following way. Suppose (1) represents the motion of a dynamical system, and $q(t)$ an external force applied to it at time t. Suppose we let $q(t)$ be a unit impulse; i.e., let $q(t)$ be 0 except for a small interval $t_0 - \delta \le t \le t_0 + \delta$, but let $\int_a^b q(t)\, dt = 1$ (Fig. 1). Then by (6), $x(t)$ is practically the same as $K(t_0, t)$; i.e., $K(s, t)$ represents the response of the system at time t to a unit impulse applied at time s. This makes it obvious why $K = 0$, $t < s$.

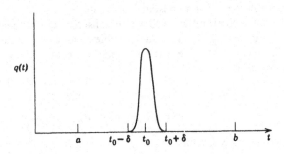

Fig. 1

If now we conceive of any force function $q(t)$ as being made up of a sum of such elementary impulses, the integral (6) says that the response of the system at time t to the force function is obtained by adding its responses to all the elementary impulses at time s, for all $s < t$.

PART C.
LINEAR SYSTEMS WITH CONSTANT COEFFICIENTS

12. Introduction. Complex Solutions

We consider the linear homogeneous system whose coefficients are real constants

$$\frac{dx_i}{dt} = \sum_{j=1}^{n} a_{ij}x_j, \qquad i = 1, 2, \cdots, n$$

i.e. $\qquad\qquad\qquad\qquad\qquad\qquad\qquad\qquad\qquad\qquad\qquad$ (1)

$$\frac{dX}{dt} = \mathscr{A} X$$

This is a special case of the general linear system of Part A. Hence the solutions of (1) are defined for all real values of t, and it is easy to verify that they possess derivatives of all orders.

The essential difference between (1) and the general linear system lies in the fact that if a vector $X(t)$ is a solution of (1), then its first derivative $X'(t)$ is also a solution. This follows immediately from the relation

$$\frac{d}{dt}(\mathscr{A}X) = \mathscr{A}\frac{dX}{dt} \tag{2}$$

$\mathscr{A}$ being constant. Hence the derivatives of any order, or any linear combination of derivatives, are also solutions. We shall always make use of the condition that $\mathscr{A}$ is a constant matrix in this way.

It will be convenient in what follows to consider not only real but also complex solutions $X(t)$ of (1). (The variable t will always be considered real.) If $X = (x_1, x_2, \cdots, x_n)$ is any complex vector, let us define the *real* and *imaginary parts* of X as

$$R(X) = \{R(x_1), R(x_2), \cdots, R(x_n)\}$$
$$I(X) = \{I(x_1(, I(x_2), \cdots, I(x_n)\} \tag{3}$$

We have immediately Lemma 1:

Lemma 1. $X(t)$ *is a solution of* (1) *if and only if both* $R(X)$ *and* $I(X)$ *are solutions of* (1).

We define *linear independence* of complex vectors and functions exactly as in the real case, except that complex constants c_i are allowed. It is clear that if a set of real vectors (or functions) is linearly independent in our previous sense it remains so if complex constants are allowed. All the statements of Part A hold without change when complex solutions of the system are admitted.

Theorem 14. From a set of n linearly independent complex solutions $X_1(t)$, $X_2(t)$, $\cdots$, $X_n(t)$ of (1), there may be constructed a set of n linearly independent real solutions of (1), which therefore constitute a fundamental system.

Proof. Let t_0 be any value of t, and consider the set of constant vectors $X_i(t_0)$, $i = 1, 2, \cdots, n$. By Theorem 2 this set is linearly independent. Hence among the $(2n)$ real vectors

$$R\{X_i(t_0)\}, \qquad I\{X_i(t_0)\}, \qquad i = 1, 2, \cdots, n \qquad (4)$$

precisely n are linearly independent. For if all the vectors (4) could be expressed in terms of some $(n - 1)$ vectors, so could $X_i(t_0)$, $i = 1$, $2, \cdots, n$. Hence among the functions

$$R\{X_i(t)\}, \qquad I\{X_i(t)\}, \qquad i = 1, 2, \cdots, n \qquad (5)$$

precisely n are linearly independent; and since by Lemma 1 they are solutions of (1) the proof is complete.

We are therefore entitled to say that (1) has been completely solved when a set of n linearly independent complex solutions is known. We shall show in the following sections that such a set may be obtained by purely algebraic means.

13. Characteristic Values and Vectors of a Matrix

In this section and the next we shall introduce the algebraic ideas necessary for the present discussion. Let $\mathscr{A} = [a_{ij}]$ be an $n \times n$ matrix of real constants. A *characteristic vector* of $\mathscr{A}$ is defined to be any real or complex vector X different from 0 such that $Y = \mathscr{A} X$ is a multiple of X

$$\mathscr{A} X = \lambda X, \quad \text{or } (\mathscr{A} - \lambda \mathscr{E})X = 0 \qquad (1)$$

where λ is some real or complex constant. λ is called a *characteristic value* of the matrix $\mathscr{A}$, and X a characteristic vector *belonging to λ*. In terms of components (1) becomes

$$
\begin{aligned}
(a_{11} - \lambda)x_1 + \qquad a_{12}x_2 + \cdots + \qquad a_{1n}x_n &= 0 \\
a_{21}x_1 + (a_{22} - \lambda)x_2 + \cdots + \qquad a_{2n}x_n &= 0 \qquad (2) \\
&\ \ \cdots \cdots \cdots \cdots \cdots \\
a_{n1}x_1 + \qquad a_{n2}x_2 + \cdots + (a_{nn} - \lambda)x_n &= 0
\end{aligned}
$$

The system (2) will admit a nontrivial solution $(x_1, x_2, \cdots, x_n)$ only for those values of λ for which the determinant of the coefficients vanishes

$$|\mathscr{A} - \lambda\mathscr{E}| = \begin{vmatrix} a_{11} - \lambda & a_{12} & \cdots & a_{1n} \\ a_{21} & a_{22} - \lambda & \cdots & a_{2n} \\ \cdot & \cdot & \cdots & \cdot \\ \cdot & \cdot & \cdots & \cdot \\ \cdot & \cdot & \cdots & \cdot \\ a_{n1} & a_{n2} & \cdots & a_{nn} - \lambda \end{vmatrix} = 0 \qquad (3)$$

A number λ will be a characteristic value of $\mathscr{A}$ if and only if it satisfies the equation (3), which is therefore known as the *characteristic equation* for $\mathscr{A}$. Since (3) is an algebraic equation of degree n, it follows that $\mathscr{A}$ cannot have more than n different characteristic values. If λ is an m-fold root of (3), it is said to be a characteristic value of *multiplicity m*. From the fundamental theorem of algebra it follows that $\mathscr{A}$ has precisely n characteristic values, each different one being counted as many times as its multiplicity.

Theorem 15. Let $X_1, X_2, \cdots, X_m$ be m characteristic vectors of $\mathscr{A}$ belonging respectively to the distinct characteristic values $\lambda_1, \lambda_2, \cdots, \lambda_m$. Then the X's are linearly independent.

Proof. The theorem is trivial for $m = 1$, since a characteristic vector cannot be 0. We give a proof by induction. Suppose the theorem true for $m = k - 1$, and let $X_1, X_2, \cdots, X_k$ be characteristic vectors of $\mathscr{A}$ belonging respectively to distinct characteristic values $\lambda_1, \lambda_2, \cdots, \lambda_k$. Suppose there exists a nontrivial linear relation

$$\sum_{i=1}^{k} c_i X_i = 0 \qquad (4)$$

Applying the matrix operator $\mathscr{A}$ to (4) we have the additional relation

$$\sum_{i=1}^{k} c_i \lambda_i X_i = 0 \qquad (5)$$

We may assume without loss of generality that at least one of the constants $c_1, c_2, \cdots, c_{k-1}$ is different from 0. Multiply (4) by λ_k and subtract (5) to eliminate $c_k \lambda_k X_k$. This gives

$$\sum_{i=1}^{k-1} c_i (\lambda_k - \lambda_i) X_i = 0 \qquad (6)$$

Since $\lambda_k - \lambda_i \neq 0$, (6) is a nontrivial relationship contrary to the induction. Hence the relation (4) cannot be nontrivial. This completes the induction.

In particular, suppose that there exist n distinct characteristic values $\lambda_1, \lambda_2, \cdots, \lambda_n$ (i.e., all the roots of the characteristic equation are simple). Then the system (2) admits of essentially only a single solution $X_i = (x_{i1}, x_{i2}, \cdots, x_{in})$ for each λ_i and the n vectors $X_1, X_2, \cdots, X_n$ so defined will be linearly independent.

14. Vectors Associated with Characteristic Values of a Matrix

We have just seen that if the characteristic equation has only simple roots, then there exists a set of n linearly independent characteristic vectors of $\mathscr{A}$. To handle the case where the characteristic equation has multiple roots, we introduce the following definition:

Definition 2. *Let λ be a root of the characteristic equation* $|\mathscr{A} - \lambda\mathscr{E}| = 0$. *Then a complex vector X different from 0 is said to be a vector associated with λ of multiplicity m, if for some integer m (and no smaller m)*

$$(\mathscr{A} - \lambda\mathscr{E})^m X = 0 \tag{1}$$

If $m = 1$, X is a characteristic vector belonging to λ. It can be shown that no vector can be associated with two different characteristic values.

We state without proof the following theorem of algebra; it follows from the general theory of elementary divisors, which is treated in any text on the algebra of matrices.

Theorem 16. *Let $\lambda_1, \lambda_2, \cdots, \lambda_k$ be the distinct roots of $|\mathscr{A} - \lambda\mathscr{E}| = 0$, with respective multiplicities $m_1, m_2, \cdots, m_k$, where $\sum_{i=1}^{k} m_i = n$. Then corresponding to each λ_i there exists a set of precisely m_i vectors X_{ij} $(j = 1, 2, \cdots, m_i)$ such that*

(1) X_{ij} $(j = 1, 2, \cdots, m_i)$ *is associated with λ_i of multiplicity $\leq m_i$.*

(2) *The set of n vectors X_{ij} $(j = 1, 2, \cdots, m_i; i = 1, 2, \cdots, k)$ is linearly independent.*

In particular, if all the multiplicities are 1 ($k = n$), this reduces to Theorem 15.

This completes the algebraic preliminaries. We shall now consider the problem of finding a fundamental system of solutions of $dX/dt = \mathscr{A}X$. Our general result will be this: if all the characteristic values of $\mathscr{A}$ are known, a fundamental system of solutions can be found by a finite number of elementary operations (i.e., solution of linear algebraic systems).

15. The Solution in the Simplest Case

In this section we shall show how a fundamental system of

$$\frac{dX}{dt} = \mathscr{A}X \tag{1}$$

may be constructed, on the assumption that the roots of the characteristic equation are all simple. In this special case we shall not need to assume Theorem 16 in its general form.

Our first two theorems (17 and 18) are independent of the nature of the roots of the characteristic equation.

Theorem 17. *Let $X(t)$ be a solution of* (1) *for all t. If, for some value t_0 of t, X is a characteristic vector of $\mathscr{A}$ belonging to the characteristic value λ*

$$\mathscr{A}X(t_0) = \lambda X(t_0) \tag{2}$$

then the same is true for every value of t

$$\mathscr{A}X(t) \equiv \lambda X(t) \tag{3}$$

Proof. Consider the function

$$Y(t) = \frac{dX(t)}{dt} - \lambda X(t) \tag{4}$$

It is a solution of (1). We may also write

$$Y(t) = \mathscr{A}X(t) - \lambda X(t) = (\mathscr{A} - \lambda\mathscr{E})X(t) \tag{5}$$

But since

$$Y(t_0) = (\mathscr{A} - \lambda\mathscr{E})X(t_0) = 0$$

by hypothesis, it follows from the uniqueness theorem that $Y(t) \equiv 0$. In view of (5) the theorem follows.

It therefore makes sense to speak of a "*characteristic solution*" $X(t)$ of (1), i.e., a solution for which (3) holds for all t. Theorem 17 then becomes: a solution $X(t)$ of (1) is a characteristic solution belonging to λ if and only if, for some value t_0 of t, $X(t_0)$ is a characteristic vector belonging to λ.

The importance of the characteristic solutions is that they may be expressed in a particularly simple form:

Theorem 18. *Corresponding to every characteristic value λ of $\mathscr{A}$, there exists at least one characteristic solution $X(t)$ of* (1), *which may be written in the form*

$$X(t) = (x_1 e^{\lambda t}, x_2 e^{\lambda t}, \cdots, x_n e^{\lambda t}) \tag{6}$$

where $(x_1, x_2, \cdots, x_n)$ is a characteristic vector belonging to λ.

Proof. Let λ and one of its characteristic vectors $X_0 = (x_1, x_2, \cdots, x_n)$ be given. Let $X(t) = \{x_1(t), x_2(t), \cdots, x_n(t)\}$ be the solution of (1) such that $X(0) = X_0$. Then by Theorem 17, $X(t)$ satisfies

$$\frac{dX(t)}{dt} = \mathscr{A} X = \lambda X(t) \tag{7}$$

Writing (7) in terms of its components, we have

$$\frac{dx_i(t)}{dt} = \lambda x_i(t), \qquad i = 1, 2, \cdots, n \tag{8}$$

Hence

$$x_i(t) = c_i e^{\lambda t}$$

and since

$$x_i(0) = c_i = x_i$$

$X(t)$ is precisely of the form (6).

From this our first main result follows immediately:

Theorem 19. *Any set of characteristic solutions of* (1) *belonging to different characteristic values is linearly independent. If in particular $\mathscr{A}$ has n different characteristic values, the corresponding set of characteristic solutions forms a fundamental system of* (1).

Proof. Taking for t any particular value t_0, the constant vectors so obtained are linearly independent by Theorem 15.

Observe that we might equally well attempt to find a solution of (1) in the form

$$X(t) = (c_1 e^{\lambda t}, c_2 e^{\lambda t}, \cdots, c_n e^{\lambda t}) \tag{9}$$

Substituting (9) into (1) the exponentials cancel, and the condition that (9) be a solution is that the c's satisfy a certain linear algebraic system. This will have a nontrivial solution if and only if λ is a characteristic value. This is essentially the method we shall use in the general case.

Observe that the fundamental system we have obtained will in general consist of complex solutions, the λ's being in general complex. By Theorem 14 this may be reduced to a real fundamental system.

16. The Solution in the General Case

We shall show in this section that there exist what we shall call *primitive solutions* of

$$\frac{dX}{dt} = \mathscr{A} X \tag{1}$$

corresponding to the vectors of $\mathscr{A}$ associated with the characteristic values λ. We shall show that these solutions may be expressed in a certain simple form; and from Theorem 16 it will follow that they constitute a fundamental system of (1).

It will be convenient to introduce the differential operator $D = d/dt$, and to write symbolic polynomials in D according to the scheme

$$D^k x \text{ means } \frac{d^k x}{dt^k}$$

$$(aD^2 + bD + c)x \text{ means } a\frac{d^2 x}{dt^2} + b\frac{dx}{dt} + cx \tag{2}$$

$$(D - a)(D - b)x \text{ means } [D^2 - (a + b)D + ab]x$$

If $X(t)$ is a solution of (1), we know (§ 12) that DX is also a solution and

$$DX = \mathscr{A}X$$

from which there follows immediately

$$D^k X = \mathscr{A}^k X$$

with k any integer. If therefore P is any polynomial with constant coefficients, we have the relation

$$P(D)X = P(\mathscr{A})X \tag{3}$$

for any solution $X(t)$ of (1).

Theorem 20. *Let $X(t)$ be a solution of* (1); *and suppose that for some value t_0 of t, $X(t_0)$ is associated with a characteristic value λ of $\mathscr{A}$ with multiplicity m*

$$(\mathscr{A} - \lambda\mathscr{E})^m X(t_0) = 0 \tag{4}$$

Then the same relation is true for all values of t

$$(\mathscr{A} - \lambda\mathscr{E})^m X(t) \equiv 0$$

and also $\tag{5}$

$$(D - \lambda)^m X(t) \equiv 0$$

Proof. It follows from (3) that

$$(\mathscr{A} - \lambda\mathscr{E})^m X(t) \equiv (D - \lambda)^m X(t) \tag{6}$$

Let us now set

$$Z(t) = (D - \lambda)^m X(t)$$

Then $Z(t)$ is a solution of (1), being a linear combination of $X(t)$ and its derivatives. But since

$$Z(t_0) = (\mathscr{A} - \lambda\mathscr{E})^m X(t_0) = 0 \tag{7}$$

by hypothesis, it follows from the uniqueness theorem that $Z(t) \equiv 0$. This completes the proof.

Let us now define a *primitive solution* of (1) associated with the characteristic value λ of multiplicity m as any nontrivial solution $X(t)$ of (1) for which there exists a characteristic value λ of $\mathscr{A}$ and an integer m such that (5) holds identically. Theorem 20 then becomes: $X(t)$ is a primitive solution of (1) if and only if there is some value t_0 of t for which $X(t_0)$ is a vector associated with the characteristic value λ of multiplicity m. The primitive solutions of (1) may, like the characteristic solutions, be written in a simple form:

Theorem 21. If $X(t)$ is a primitive solution of (1) associated with the characteristic value λ of multiplicity m

$$(\mathscr{A} - \lambda\mathscr{E})^m X(t) \equiv 0 \tag{8}$$

then $X(t)$ is of the form

$$X(t) = \{P_1(t)e^{\lambda t}, P_2(t)e^{\lambda t}, \cdots, P_n(t)e^{\lambda t}\} \tag{9}$$

where $P_i(t)$ $(i = 1, 2, \cdots, n)$ are polynomials in t of degree $\leq m - 1$.

Proof. By Theorem 20 we have

$$(D - \lambda)^m X(t) \equiv 0 \tag{10}$$

Writing (10) in scalar form, where

$$X(t) = \{x_1(t), x_2(t), \cdots, x_n(t)\}$$

we have

$$(D - \lambda)^m x_1(t) \equiv 0, \qquad i = 1, 2, \cdots, n \tag{11}$$

Now for any complex function $x(t)$ with continuous derivatives of mth order, we observe that

$$(D - \lambda)^m [e^{\lambda t} x(t)] = e^{\lambda t} D^m [x(t)] \tag{12}$$

with λ any constant. Equation (12) is trivial for $m = 0$; assuming (12) for $m = k - 1$, we have

$$(D - \lambda)^k [e^{\lambda t} x(t)] = (D - \lambda)(D - \lambda)^{k-1} [e^{\lambda t} x(t)]$$

$$= (D - \lambda)[e^{\lambda t} D^{k-1}(x)] = e^{\lambda t} D^k(x)$$

Hence (12) is proved by induction. Now since (11) may be written as

$$(D - \lambda)^m [e^{\lambda t} \{e^{-\lambda t} x_i(t)\}] \equiv 0$$

by (12) we obtain

$$e^{\lambda t} D^m \{ e^{-\lambda t} x(t) \} = 0 \qquad (13)$$

Canceling $e^{\lambda t}$ in (13) and integrating, we have

$$x_i(t) = e^{\lambda t} P_i(t), \qquad i = 1, 2, \cdots, n \qquad (14)$$

where $P_i(t)$ $(i = 1, 2, \cdots, n)$ is a polynomial of degree $\leq m - 1$. This is precisely (9), and the proof is complete.

Our general result now follows:

Theorem 22. A fundamental system of solutions of (1) *of the form* (9) *always exists; and if all the characteristic values* λ_i *of the matrix* $\mathscr{A}$ *are known, this fundamental system may be obtained by a finite number of elementary operations.*

Proof. The existence is trivial; for by Theorem 16 there exists a linearly independent set of n vectors X_{i0} of $\mathscr{A}$ associated with its characteristic values. If we define solutions $X_i(t)$ of (1) satisfying the initial conditions $X_i(t_0) = X_{i0}$, then these will be linearly independent and of the form (9).

Let new $\lambda_1, \lambda_2, \cdots, \lambda_k$ be the characteristic values of $\mathscr{A}$, with respective multiplicities $m_1, m_2, \cdots, m_k$. Let λ be any one of them, with multiplicity m. We shall try to satisfy (1) by functions of the form

$$X(t) = \{ P_{i1}(t) e^{\lambda t}, P_{i2}(t) e^{\lambda t}, \cdots, P_{in}(t) e^{\lambda t} \} \qquad (15)$$

where $P_{ij}(t)$ are polynomials of the ith order with undetermined coefficients, where $i \leq m - 1$. First take $i = 0$, and substitute (15) in (1). The exponentials will all cancel and we get a set of linear algebraic equations in the undetermined coefficients of the P's. This system may have no, or one, or several nontrivial linearly independent solutions. Performing the same process for $i = 1$, we equate coefficients of t and constants to obtain $2n$ linear equations in the $2n$ undetermined coefficients. This system again will have a certain number of linearly independent solutions. Proceeding in this way up to $i = m - 1$, our theory proves that we shall obtain at least m solutions of which precisely m will be linearly independent. Then if we do the same for the other λ's, we shall have constructed a fundamental system. Thus we have a perfectly definite method by which a fundamental system of (1) may be obtained, and the process involves nothing more difficult than the solution of linear algebraic systems. This completes the proof.

It is easily seen that the process just described is equivalent to finding n linearly independent vectors associated with the characteristic values of $\mathscr{A}$. Now Theorem 16 may be made considerably more precise, and the process of finding the required vectors will be shortened thereby. This, however, would involve more algebraic theory than we can go into here.

17. Homogeneous Equation of nth Order

We consider now the equation

$$\frac{d^n x}{dt^n} + \sum_{i=1}^{n} a_i \frac{d^{n-i} x}{dt^{n-i}} = 0 \tag{1}$$

where the a_i's are constants. We could reduce (1) to a linear system and treat it by the methods of § 16; it is, however, much simpler to treat (1) separately, and in addition we shall not need to assume Theorem 16. The relation to the methods of § 16 will be very obvious.

Equation (1) may be written in the form

$$P(D)x = 0 \tag{2}$$

where $P(D)$ is a certain polynomial. The *characteristic equation* of (1) is

$$P(\lambda) = 0 \tag{3}$$

Let $\lambda_1, \lambda_2, \cdots, \lambda_k$ be its roots, where

$$P(\lambda) = (\lambda - \lambda_1)^{m_1}(\lambda - \lambda_2)^{m_2} \cdots (\lambda - \lambda_k)^{m_k} \tag{4}$$

and

$$\sum_{i=1}^{k} m_i = n$$

We may apply the same factorization to $P(D)$ and write (2) as

$$(D - \lambda_1)^{m_1}(D - \lambda_2)^{m_2} \cdots (D - \lambda_k)^{m_k} x = 0 \tag{5}$$

Now it is clear that any function $x(t)$ satisfying

$$(D - \lambda_i)^{m_i} x = 0 \tag{6}$$

will also be a solution of (5). We have already integrated an equation of the form (6) in Theorem 21; and the most general solution of (6) is

$$x(t) = P_{m_i-1}(t)e^{\lambda_i t} \tag{7}$$

where $P_k(t)$ is a polynomial of degree k. Hence in particular the n functions

$$x_{ij}(t) = t^{j-1}e^{\lambda_i t}, \quad j = 1, 2, \cdots, m_i, \quad i = 1, 2, \cdots, k \tag{8}$$

are all solutions of (1).

Theorem 23. *The n solutions (8) of (1) are linearly independent and hence form a fundamental system of (1).*

Proof. Any linear combination of the functions (8) can be written in the form

$$\sum_{i=1}^{k} P_i(t)e^{\lambda_i t}$$

where $\lambda_i \neq \lambda_j$ if $i \neq j$, and $P_i(t)$ are polynomials in t. It will suffice therefore to prove that if

$$\sum_{i=1}^{k} P_i(t)e^{\lambda_i t} \equiv 0 \tag{9}$$

then

$$P_i(t) \equiv 0, \qquad i = 1, 2, \cdots, k$$

We give a proof by induction. The theorem is trivial for $k = 1$. Suppose it true for $k = m - 1$. Consider then the identity

$$\sum_{i=1}^{m} P_i(t)e^{\lambda_i t} \equiv 0, \qquad \lambda_i \neq \lambda_j \text{ if } i \neq j \tag{10}$$

Let us divide by $e^{\lambda_m t}$. Then (10) becomes

$$\sum_{i=1}^{m-1} P_i(t)e^{(\lambda_i - \lambda_m)t} + P_m(t) \equiv 0 \tag{11}$$

where $\lambda_i - \lambda_m \neq 0$, $i = 1, 2, \cdots, m - 1$. Differentiating (11) we have

$$\sum_{i=1}^{m-1} Q_i(t)e^{(\lambda_i - \lambda_m)t} + P_m'(t) \equiv 0$$

where $Q_i(t) = P_i'(t) + (\lambda_i - \lambda_m)P_i(t)$. Hence the Q's are polynomials of the same degree as the P's. After differentiating a suitable number of times, the term in $P_m(t)$ will disappear and we have

$$\sum_{i=1}^{m-1} R_i(t)e^{(\lambda_i - \lambda_m)t} \equiv 0 \tag{12}$$

where by the same argument the R_i's are polynomials of the same degrees as the P's. But since no two of the exponents $(\lambda_i - \lambda_m)$ are equal, our induction applies, and all the R's are identically 0. Therefore the P's, which were of the same degree as the R's, also are identically 0. The same of course holds for $P_m(t)$; and this completes the induction.

18. Applications

As a simple example of our methods, we consider the equation

$$\frac{d^2x}{dt^2} + 2k\frac{dx}{dt} + \omega^2 x = F(t) \tag{1}$$

where k and ω are real constants. Equation (1) may be interpreted as the expression of Newton's law for a particle with one degree of freedom, where

(1) $\omega^2 x$ represents a restoring force towards the position $x = 0$ proportional to the displacement.

(2) $2k(dx/dt)$ (for $k > 0$) represents a damping force proportional to the velocity.

(3) $F(t)$ is a variable external force.

If $F(t) \equiv 0$, (1) becomes homogeneous and the characteristic equation is

$$\lambda^2 + 2k\lambda + \omega^2 = 0 \tag{2}$$

The cases where the roots of (2) are equal or one is 0 are of comparatively little physical interest. Otherwise, if $k^2 > \omega^2$, the roots are real and distinct

$$\lambda_1 = -k + \sqrt{k^2 - \omega^2}, \quad \lambda_2 = -k - \sqrt{k^2 - \omega^2}$$

and the general solution is

$$x = c_1 e^{\lambda_1 t} + c_2 e^{\lambda_2 t}$$

If $k^2 < \omega^2$, the roots are complex

$$\lambda = -k \pm i\beta, \quad \text{where } \beta^2 = \omega^2 - k^2$$

and the general solution is

$$x = e^{-kt}(c_1 \cos \beta t + c_2 \sin \beta t) \tag{3}$$

By means of these results, we shall obtain the solution of the non-homogeneous equation (1), using both of the methods we have developed. First, using the method of variation of constants, we know that a particular solution of (1) may be written in the form

$$x = c_1(t)x_1 + c_2(t)x_2 \tag{4}$$

where x_1 and x_2 are a fundamental system of the homogeneous equation

$$x_1 = e^{-kt} \cos \beta t, \quad x_2 = e^{-kt} \sin \beta t$$

and $c_1(t)$ and $c_2(t)$ satisfy

$$\begin{aligned} c_1'(t)x_1 + c_2'(t)x_2 &= 0 \\ c_1'(t)x_1' + c_2'(t)x_2' &= F(t) \end{aligned} \tag{5}$$

Solving (5) we have

$$c_1'(t) = -\frac{x_2 F(t)}{W(x_1, x_2)} = -\frac{1}{\beta} e^{kt} \sin \beta t F(t)$$

$$c_2'(t) = \frac{x_1 F(t)}{W(x_1, x_2)} = \frac{1}{\beta} e^{kt} \cos \beta t F(t) \tag{6}$$

Let us integrate (6) and substitute in (4). Then after some simplifications we obtain the particular solution of (1)

$$x(t) = \frac{1}{\beta}\int_{t_0}^{t} e^{-k(t-s)} \sin \beta(t - s) F(s)\, ds \tag{7}$$

Let us now construct Green's function $K(s, t)$ for (1). We know

$$K(s, t) \equiv 0 \qquad \text{for } t < s$$

$$K(s, t) = e^{-kt}(c_1 \cos \beta t + c_2 \sin \beta t) \qquad \text{for } t > s$$

$$[K(s, t)]_{t=s+0} = 0, \qquad [K_t'(s, t)]_{t=s+0} = -1$$

The constants c_1 and c_2 are uniquely defined by the initial conditions

$$c_1 = \frac{1}{\beta}(e^{ks} \sin \beta s), \qquad c_2 = \frac{1}{-\beta}(e^{ks} \cos \beta s)$$

Hence after some simplifications $K(s, t)$ is seen to be

$$K(s, t) \equiv 0, \qquad t \leq s$$

$$K(s, t) = -\frac{1}{\beta} e^{-k(t-s)} \sin \beta(t - s), \qquad t \geq s \tag{8}$$

Now the solution $x(t)$ of (1) such that

$$x(t_0) = 0, \qquad x'(t_0) = 0$$

is known to be given by

$$x(t) = -\int_{t_0}^{t} K(s, t) F(s)\, ds \tag{9}$$

But (9) is exactly the same as (7); thus our two methods have produced precisely the same solution.

Observe that $K(s, t)$ is here a function of $t - s$ alone. This is true in general for the Green's function of any nth order equation

$$L(x) + F(t) = 0$$

provided the coefficients in L are constant.   For if $x(t)$ is a solution of $L(x) = 0$, so is $x(t - t_0)$.   Hence we can write

$$K(s, t) = K(s - s, t - s) = H(t - s) \tag{10}$$

In terms of our physical interpretation, $H(t)$ represents the response of the system at any time to a unit impulse t units of time previous.   In our case

$$H(t) = -\frac{1}{\beta} e^{-kt} \sin \beta t$$

Thus the magnitude of the response either approaches 0 or becomes unbounded as t increases, depending upon the sign of k.

4

Singularities of
an Autonomous System

PART A.
INTRODUCTION

1. Characteristic Curves

We consider in this chapter a second-order system of the form

$$\frac{dx}{dt} = P(x, y), \qquad \frac{dy}{dt} = Q(x, y) \tag{1}$$

Such a system, in which the independent variable t does not appear explicitly in the functions on the right-hand side, is called *autonomous*. Let us assume that $P(x, y)$ and $Q(x, y)$ are defined in some domain D of the x-y plane, and satisfy a Lipschitz condition on both x and y in some neighborhood of every point of D. Then by the theory of Chapter 2, if t_0 is any number and (x_0, y_0) any point of D, there exists a unique solution of (1)

$$x = x(t), \qquad y = y(t)$$

satisfying $\qquad\qquad\qquad\qquad\qquad\qquad\qquad\qquad\qquad\qquad$ (2)

$$x(t_0) = x_0, \qquad y(t_0) = y_0$$

Furthermore we know that either

(1) The solutions (2) are defined for all real values of t; or

(2) If the solutions (2) are not defined for $t > t_1$ (say), then either

(a) As $t \to t_1 - 0$, the point $(x(t), y(t))$ approaches the boundary of D;

or if D is unbounded, possibly

(b) As $t \to t_1 - 0$, either $x(t)$ or $y(t)$ or both become unbounded.

Analogous alternatives arise if t is not defined for $t <$ some number t_2.

In any subdomain D' of D in which $P(x, y)$ does not vanish, we may write (1) in the form

$$\frac{dy}{dx} = \frac{Q(x, y)}{P(x, y)} \tag{3}$$

This is simply a restricted direction field of the type studied in Chapter 1; whence we know that through any point of D' there passes a unique integral curve of (3). Essentially the same theorem is true for the more general system (1), as we shall see immediately.

Let us consider a solution (2) of (1), with the restriction that $x(t)$ and $y(t)$ are not both constants; this case will be treated later. Then (2) defines a curve with continuously turning tangent, which we call a *characteristic curve* or simply a *characteristic* of (1). We have immediately:

Theorem 1. *Through any point of D there passes at most one characteristic.*

For let $(x_1(t), y_1(t))$, $(x_2(t), y_2(t))$ be solutions of (1) representing characteristics C_1 and C_2 of (1), with the common point (x_0, y_0). Then there exist values t_1 and t_2 of t such that

$$x_1(t_1) = x_2(t_2) = x_0, \qquad y_1(t_1) = y_2(t_2) = y_0$$

Consider the pair of functions

$$(x_1(t + t_1 - t_2), y_1(t + t_1 - t_2)) \tag{4}$$

Then (4) is a solution of (1), since P and Q do not contain t explicitly. But since for $t = t_2$ we have

$$x_1(t + t_1 - t_2) = x_2(t), \qquad y_1(t + t_1 - t_2) = y_2(t) \tag{5}$$

it follows from the uniqueness theorem that (5) holds identically in t. Since (4) represents the characteristic C_1, C_1 and C_2 are identical, and the theorem follows.

Of course, the theorem is not true in general unless (1) is autonomous, as may easily be shown by examples.

It should be understood that the terms *characteristic* and *solution* are not synonymous; a characteristic is a curve in D, which may be represented parametrically by more than one solution of (1). For example, if $(x(t), y(t))$ represents a certain characteristic C, then for any t_0, $(x(t - t_0), y(t - t_0))$ is a different solution also representing C. It is clear from Theorem 1 that *all* solutions representing C may be written in this form by a suitable choice of t_0. Hence it follows that for every solution of (1) representing C, the direction of increase of t along C is the same. Therefore a characteristic is a *directed* curve; in our figures we shall represent the direction of increasing t by an arrow.

If there exists a solution of the form

$$x(t) \equiv x_0, \qquad y(t) \equiv y_0 \tag{6}$$

it is clear that no characteristic can pass through (x_0, y_0). The point (x_0, y_0) is said to be a *singularity* of the system. The condition that (6) be a solution of (1) is that

$$P(x_0, y_0) = 0, \qquad Q(x_0, y_0) = 0 \tag{7}$$

Conversely, if (7) holds for a point (x_0, y_0), it is obviously a singularity. Therefore the singularities of (1) are the points at which P and Q both vanish; they are the only points of D through which no characteristic passes. All other points of D are said to be *regular*.

We can interpret the situation kinematically in the following way. Consider the *vector field* defined by

$$V(x, y) = (P(x, y), Q(x, y)) \tag{8}$$

i.e., at (x, y) the vector $V(x, y)$ has a horizontal component $P(x, y)$ and a vertical component $Q(x, y)$. Then (1) defines the motion of a particle (x, y) in the plane by the condition that its *velocity* be at every point equal to the vector $V(x, y)$. Then the characteristics of (1) represent the *trajectories* of the particle, i.e., those fixed paths along which it must move, independently of its starting point. The singularities represent its points of *equilibrium*, i.e., those points at which it is at rest. In this chapter we shall consider the *type of equilibrium* defined by certain simple kinds of singularities of the system (1).

Our theorem implies as a corollary that no characteristic can cross itself. For if C: $(x(t), y(t))$ is a characteristic and there exist two values t_0 and t_1 of t such that

$$x(t_0) = x(t_1), \qquad y(t_0) = y(t_1) \tag{9}$$

it follows that for all t

$$x(t_1 + t) \equiv x(t_0 + t), \qquad y(t_1 + t) \equiv y(t_0 + t)$$

That is to say, C is a *closed* curve. C will have a *period h*, where h is the least number such that

$$x(t + h) \equiv x(t), \qquad y(t + h) \equiv y(t)$$

The existence of closed characteristics will prove to be of fundamental importance in the theory of characteristics in the large of Chapter 5.

2. Isolated Singularities

From the results of § 1, we know concerning any characteristic C: $(x(t), y(t))$ that either:

(1) $x(t)$ and $y(t)$ are defined for all real values of t; or

(2) If $x(t)$ and $y(t)$ are undefined for $t \geq t_0$, then as $t \to t_0 - 0$ the point P: $(x(t), y(t))$ either approaches the boundary of D or becomes unbounded. It is clear that these results are independent of the particular solution $(x(t), y(t))$ used to represent C.

Let us now define a *half-characteristic* as the set of points of a characteristic C such that, for some representation $(x(t), y(t))$ of C, $t \geq$ some number t_0. Then we have from the above a lemma:

Lemma 1. *If a half-characteristic is contained within a closed bounded region R in D, its representing solution is defined for all $t \geq t_0$. This lemma will be of constant use in the next chapter.*

Consider now an *isolated singularity* (x_0, y_0) of

$$\frac{dx}{dt} = P(x, y), \qquad \frac{dy}{dt} = Q(x, y) \tag{1}$$

i.e., suppose there exists a circle $(x - x_0)^2 + (y - y_0)^2 \leq \rho^2$ containing no singularity of (1) except (x_0, y_0). For convenience, take (x_0, y_0) to be the point $(0, 0)$. Suppose there exists a half-characteristic C: $(x(t), y(t))$ lying in some closed bounded region R that contains $(0, 0)$ as its interior point. Then $x(t)$ and $y(t)$ are defined for all $t \geq t_0$. We say that C *approaches the singularity* $(0, 0)$ if

$$\lim_{t \to \infty} x(t) = 0, \qquad \lim_{t \to \infty} y(t) = 0$$

or equivalently if

$$\lim_{t \to \infty} [x^2(t) + y^2(t)]^{\frac{1}{2}} = 0 \tag{2}$$

In terms of point-set theory, the derived set C' of C is the single point $(0, 0)$. It is hardly necessary to state that the fulfillment of (2) does not depend upon the particular solution $(x(t), y(t))$ that we happen to be using. In general, wherever we define properties of a characteristic by means of a representing solution, it will be understood that the property is independent of the solution chosen.

If a half-characteristic C approaches the singularity, we shall say further that it *enters* the singularity if the radius vector from $(0, 0)$ to a point of C has a limiting direction as $t \to \infty$; i.e., if

$$\lim_{t \to \infty} \frac{y(t)}{x(t)} \tag{3}$$

exists or diverges properly to $\pm \infty$. A sufficient condition that C enter the singularity is that the direction of the tangent to C at (x, y) have a limiting position as $t \to \infty$; for by l'Hospital's rule, if $\lim_{t \to \infty} \dfrac{y'(t)}{x'(t)}$ exists (or $= \pm \infty$) then $\lim_{t \to \infty} \dfrac{y(t)}{x(t)}$ exists and has the same value. For most singularities in which we are interested, this is a necessary condition as well (cf. § 10).

In the preceding definitions, we may also consider limits as $t \to -\infty$, and consider the behavior of *negative* half-characteristics in the neighborhood of a singularity. We shall in fact for mathematical purposes frequently replace t by $-t$; though not forgetting that for the physicist t is generally time.

We have already mentioned that a singularity of (1) may be represented as a point of equilibrium of a kinematical system. We wish to formulate precisely the idea of *stability* of such an equilibrium. This may be done in more than one way; the following definition is due to Liapunov:

Definition 1. *Let P: (x_0, y_0) be an isolated singularity of* (1). *P is said to represent a state of stable equilibrium if corresponding to every number $\epsilon > 0$ there exists a number $\delta > 0$ with the following property. Let C: $(x(t), y(t))$ be any characteristic satisfying*

$$\sqrt{(x(t_0) - x_0)^2 + (y(t_0) - y_0)^2} < \delta \tag{4}$$

for some value t_0 of t. Then C is defined for all $t \geq t_0$ and

$$\sqrt{(x(t) - x_0)^2 + (y(t) - y_0)^2} < \epsilon, \qquad t_0 \leq t < \infty \tag{5}$$

Roughly, we may say the condition is this: once a characteristic gets very close to the singularity, it stays close to the singularity. If every characteristic passing through some point of a certain circle

$$(x - x_0)^2 + (y - y_0)^2 \leq \rho^2$$

approaches P, then clearly the equilibrium is stable.

In this chapter we shall classify the singularities of the system (1) in terms of the *local behavior* of the characteristics near them. To do this we shall first consider the case where (1) is a *linear* system. We shall then prove that in the general case we may under certain conditions replace $P(x, y)$ and $Q(x, y)$ by the linear terms of their Taylor expansions about (x_0, y_0), without any qualitative change in the local geometric behavior of the characteristics.

PART B.
SINGULARITIES OF A LINEAR SYSTEM

3. General

Consider the linear system

$$\frac{dx}{dt} = ax + by, \qquad \frac{dy}{dt} = cx + dy \tag{1}$$

where a, b, c, and d are constants. The system (1) has a singularity at $(0, 0)$; let us assume that it has no other singularity; i.e., that

$$\begin{vmatrix} a & b \\ c & d \end{vmatrix} \neq 0$$

The characteristic equation of (1) is

$$\begin{vmatrix} a - \lambda & b \\ c & d - \lambda \end{vmatrix} = \lambda^2 - (a + d)\lambda + (ad - bc) = 0 \tag{2}$$

Observe that 0 cannot be a characteristic value. We know how to determine the solutions of (1) from the roots λ_1, λ_2 of (2). We shall prove that the nature of the singularity $(0, 0)$ of (1) is completely determined by the values of λ_1 and λ_2.

It will frequently be convenient to introduce new variables $\tilde{x}, \tilde{y}$ by the affine transformation

$$\begin{aligned} \tilde{x} &= px + qy \\ \tilde{y} &= rx + sy \end{aligned} \qquad \text{where } \begin{vmatrix} p & q \\ r & s \end{vmatrix} \neq 0 \tag{3}$$

We shall then obtain a new system of the form

$$\begin{aligned} \frac{d\tilde{x}}{dt} &= A\tilde{x} + B\tilde{y} \\ \frac{d\tilde{y}}{dt} &= C\tilde{x} + D\tilde{y} \end{aligned} \qquad \text{where } \begin{vmatrix} A & B \\ C & D \end{vmatrix} \neq 0 \tag{4}$$

It is well known that (4) will have the same characteristic values as (1). Indeed, in terms of matrices,

$$\begin{bmatrix} A & B \\ C & D \end{bmatrix} = \begin{bmatrix} p & q \\ r & s \end{bmatrix} \begin{bmatrix} a & b \\ c & d \end{bmatrix} \begin{bmatrix} p & q \\ r & s \end{bmatrix}^{-1}$$

It is easy to see that all the properties of characteristics in which we are interested (their approaching or entering a singularity, their being closed or bounded, etc.) are invariant under affine transformation. The transformation will only have the effect of somewhat distorting the shape of the characteristics. We shall in every case choose the constants of (3) so that (4) will have as simple a form as possible; (4) will then be a *canonical form* of (1).

We shall proceed to classify the singularities of (1) according to the following scheme:

Case I. λ_1 and λ_2 are real and unequal.

(a) λ_1 and λ_2 have the same sign.

(b) λ_1 and λ_2 have different signs.

Case II. There is a double root.

(a) The rank of $\begin{pmatrix} a - \lambda & b \\ c & d - \lambda \end{pmatrix}$ is 0.

(b) The rank of $\begin{pmatrix} a - \lambda & b \\ c & d - \lambda \end{pmatrix}$ is 1.

Case III. λ_1 and λ_2 are conjugate complex.

(a) The real part of the λ's is not 0.

(b) λ_1 and λ_2 are purely imaginary.

4. Nodal Points

We consider first Case I(a), where λ_1 and λ_2 are real, unequal, and of the same sign. For definiteness let us take $\lambda_1 < \lambda_2 < 0$. (The case where the λ's are positive will be obtained by substituting $-t$ for t.) We shall first consider the general system

$$\frac{dx}{dt} = ax + by, \qquad \frac{dy}{dt} = cx + dy \qquad (1)$$

without reducing to canonical form. Then, as may be verified by substitution into (1), two particular solutions are given by

$$\begin{aligned} x_1 &= Ae^{\lambda_1 t}, & y_1 &= Be^{\lambda_1 t} \\ x_2 &= Ce^{\lambda_2 t}, & y_2 &= De^{\lambda_2 t} \end{aligned} \qquad (2)$$

where A, B, C, D are definite constants and $\begin{vmatrix} A & B \\ C & D \end{vmatrix} \neq 0$. The characteristics represented by (2) are half-lines from the origin with different slopes. If we multiply the solutions (2) by -1, we have two more half-lines and the four characteristics so defined form, together with the origin, two straight lines.

The most general solution may be written in the form

$$x = c_1 A e^{\lambda_1 t} + c_2 C e^{\lambda_2 t}$$
$$y = c_1 B e^{\lambda_1 t} + c_2 D e^{\lambda_2 t} \tag{3}$$

with c_1 and c_2 arbitrary. Clearly, all the characteristics approach the origin, and the rectilinear ones enter it with slopes B/A and D/C. Furthermore, if in (2) we take $c_2 \neq 0$, we have

$$\lim_{t \to \infty} \frac{y(t)}{x(t)} = \lim_{t \to \infty} \left(\frac{D + \dfrac{c_1}{c_2} B e^{(\lambda_1 - \lambda_2)t}}{C + \dfrac{c_1}{c_2} A e^{(\lambda_1 - \lambda_2)t}} \right) = \frac{D}{C} \tag{4}$$

Therefore all the characteristics enter the origin, and all the nonrectilinear characteristics are tangent to a certain one of the rectilinear characteristics at the origin. We have then Fig. 1.

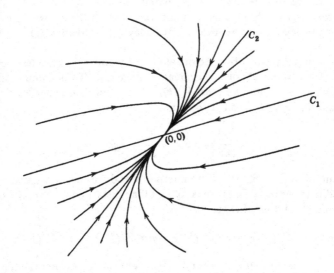

Fig. 1

It is easy to verify that for any nonrectilinear characteristic (3)

$$\lim_{t \to -\infty} \frac{y(t)}{x(t)} = \frac{B}{A}$$

In (3) take c_1 and c_2 positive. Then every ray from 0 between the rectilinear characteristics C_1 and C_2 meets (3) once and once only. Now the curve obtained from (3) by a *similarity transformation*

$$\tilde{x}(t) = kx(t), \qquad \tilde{y}(t) = ky(t) \tag{5}$$

with k any positive constant, is also a solution of (1). Hence by expanding and contracting any characteristic C of the form (3) we obtain *all* the characteristics in the region between C_1 and C_2, and similarly for the other three regions of the plane. This will in general be true for linear systems: one nonrectilinear characteristic defines *all* the characteristics in a certain region by means of the similarity transformation.

The singularity represented by Fig. 1 is called a *nodal point*; more generally, a nodal point is a singularity entered by all characteristics that start sufficiently close to it. In the present case there are four rectilinear characteristics making up (together with the origin) two straight lines; all the other characteristics are tangent to one of these lines at the origin. The equilibrium is certainly stable.

If now we replace t by $-t$, we obtain the case $\lambda_1 > \lambda_2 > 0$. Then our work will go through as before, except that all the arrows in Fig. 1 will point *away* from the origin. We may call this an *unstable* nodal point.

Let us now take any two vectors along the rectilinear characteristics as unit vectors for an oblique co-ordinate system $\tilde{x}$, $\tilde{y}$. To be more explicit, let $x = A\tilde{x} + C\tilde{y}$ and $y = B\tilde{x} + D\tilde{y}$. Then $\tilde{x}$ and $\tilde{y}$ are an affine transformation of x and y, and (1) becomes

$$\frac{d\tilde{x}}{dt} = \lambda_1 \tilde{x}, \qquad \frac{d\tilde{y}}{dt} = \lambda_2 \tilde{y} \tag{6}$$

the canonical form of (1). If we interpret $\tilde{x}$ and $\tilde{y}$ as being perpendicular co-ordinates, we are simply making an affine distortion of Fig. 1. Now the general solution of (6) is

$$\tilde{x} = c_1 e^{\lambda_1 t}, \qquad \tilde{y} = c_2 e^{\lambda_2 t} \tag{7}$$

where c_1 and c_2 are arbitrary. The four semiaxes are rectilinear characteristics, and every nonrectilinear characteristic is tangent to the y-axis at the origin. Finally, from any nonrectilinear characteristic we may "fill up" the first quadrant by a similarity transformation. We then have Fig. 2.

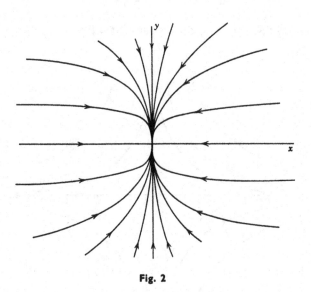

Fig. 2

Evidently, it is easier to work with the canonical forms of (1). We shall do this in the future, and merely indicate what will be the changes in our figures for the general case.

5. Saddle Points

We consider now the case where λ_1 and λ_2 are of opposite sign, say $\lambda_2 < 0 < \lambda_1$. Then as in § 4 we may consider our system in the canonical form

$$\frac{dx}{dt} = \lambda_1 x, \qquad \frac{dy}{dt} = \lambda_2 y \tag{1}$$

The general solution is

$$x = c_1 e^{\lambda_1 t}, \qquad y = c_2 e^{\lambda_2 t} \tag{2}$$

The four semiaxes are all characteristics, directed as in Fig. 3. If in (2) we take $c_1 = c_2 = 1$, for the characteristic so defined we have:

$$\text{as } t \to \infty, \ x \to \infty, \ y \to 0$$

$$\text{as } t \to -\infty, \ x \to 0, \ y \to \infty$$

We may "fill up" the first quadrant and obtain Fig. 3.

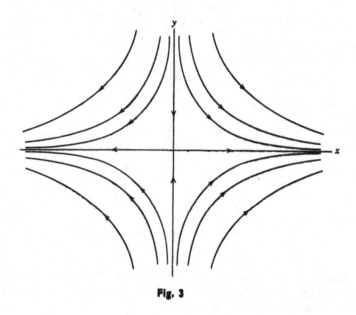

Fig. 3

We thus have four rectilinear characteristics, asymptotes to all the others. If a half-characteristic starts on the y-axis, it will enter the origin along the y-axis; in any other case it will approach the x-axis at $x = \pm\infty$. Such a singularity is known as a *saddle point*; it represents an unstable equilibrium. In the general case, the rectilinear characteristics will not be perpendicular, but the configuration will remain unchanged.

6. Degenerate Nodal Points

Although the case where the characteristic equation has a double root λ is of less importance, we include it for the sake of completeness. We distinguish between two subcases.

(a) Suppose the rank of $\begin{pmatrix} a - \lambda & b \\ c & d - \lambda \end{pmatrix}$ is 0. Then $b = c = 0$, $a = d = \lambda$, and the system is already in a canonical form

$$\frac{dx}{dt} = \lambda x, \qquad \frac{dy}{dt} = \lambda y \tag{1}$$

The solutions are all rectilinear characteristics

$$x = c_1 e^{\lambda t}, \qquad y = c_2 e^{\lambda t}$$

which (if $\lambda < 0$) all enter the origin as $t \to \infty$; see Fig. 4.

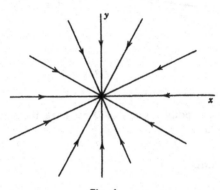

Fig. 4

This is a very simple nodal point. If $\lambda > 0$ we have the same picture except that all the arrows point *away* from the origin.

(b) Suppose the rank of $\begin{pmatrix} a - \lambda & b \\ c & d - \lambda \end{pmatrix}$ is 1. Then there exists essentially a single solution of the form

$$x = A e^{\lambda t}, \qquad y = B e^{\lambda t} \tag{2}$$

Now it is known from the further theory of linear systems with constant coefficients (which we have not considered) that there is another solution of the form

$$x = (At + C)e^{\lambda t}, \qquad y = (Bt + D)e^{\lambda t} \tag{3}$$

From (2) it appears that there are two rectilinear characteristics forming, together with the origin, a single line. Any nonrectilinear characteristic may be written as a multiple of (3), and (if $\lambda < 0$)

$$\lim_{t \to \infty} \frac{y}{x} = \frac{B}{A}$$

Therefore all the characteristics approach the origin tangent to one of the rectilinear characteristics. Since the characteristics are symmetric in the origin, we must have a picture of the form shown in Fig. 5. This is a

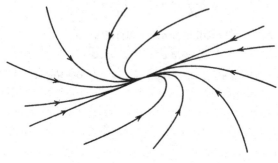

Fig. 5

degenerate nodal point. For $\lambda > 0$ we have the same figure but with arrows reversed.

7. Vortex Points

Suppose now the characteristic values are complex. We may then set $\lambda_1 = \alpha + i\beta$, $\lambda_2 = \alpha - i\beta$, where $\alpha = \frac{1}{2}(a + d)$, $\beta^2 = -bc - [\frac{1}{2}(a - d)]^2 > 0$. By a suitable real affine transformation we may consider our system in the canonical form

$$\frac{dx}{dt} = \alpha x - \beta y, \qquad \frac{dy}{dt} = \beta x + \alpha y \qquad (1)$$

We consider first the case when $\alpha = 0$, i.e., λ_1 and λ_2 are purely imaginary. Then a solution of (1) is

$$x = \cos \beta t, \qquad y = \sin \beta t \qquad (2)$$

which represents a circle about the origin. By a similarity transformation it appears that all the characteristics are circles with center at the origin. The picture in the general case will be an affine distortion of this, i.e., a family of ellipses with center at the origin, as shown in Fig. 6.

Such a singularity is called a *vortex point*; in general a vortex point is one for which every characteristic in some neighborhood of the singularity is closed and surrounds the singularity. Although no characteristic approaches the origin, it is clear from our definition that such a singularity is stable.

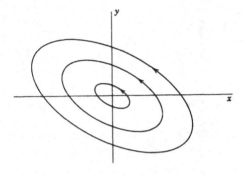

Fig. 6

Let the system again be written in the general form

$$\frac{dx}{dt} = ax + by, \qquad \frac{dy}{dt} = cx + dy \tag{3}$$

Then $bc < 0$. Since for $y = 0$, $dy/dt = cx$, it follows that the direction of rotation of the characteristics is that of the sign of c, or the negative of the sign of b (counterclockwise rotation being considered positive).

8. Spiral Points

Suppose finally the characteristic values are complex but not purely imaginary (i.e., $\alpha \neq 0$). Then in the notation of § 7 there is a solution of the form

$$x = e^{\alpha t} \cos \beta t, \qquad y = e^{\alpha t} \sin \beta t \tag{1}$$

This is a logarithmic spiral about the origin, from which we may obtain all other characteristics by similarity. Such a singularity is known as a *spiral* or *focal point*. If $\alpha < 0$, it is clear that all the characteristics approach the origin as $t \to \infty$, but none enters it. The equilibrium is stable.

If $\alpha > 0$, all the spirals run away from the origin towards ∞. For a system in the form of (3) of § 7, the rule for direction of rotation is the same. The picture in the general case will therefore be that of a family of affinely distorted logarithmic spirals, with the four possible configurations illustrated in Fig. 7.

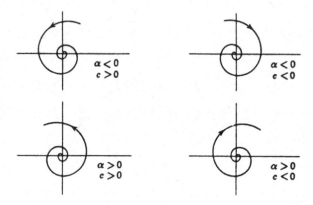

Fig. 7

We have drawn only one spiral in each case. The reader will have to imagine the rest of the family of spirals in between the turns of the first one.

9. Conclusion. Dynamical Interpretation

We may then summarize our results in the following way:
(1) λ_1, λ_2 real and of the same sign: nodal point.
(2) λ_1, λ_2 real and of opposite sign: saddle point.
(3) λ_1, λ_2 purely imaginary: vortex point.
(4) λ_1, λ_2 complex: spiral point.

As regards stability, by considering separate cases we arrive at the result: a necessary and sufficient condition that the equilibrium represented by the singularity $(0, 0)$ of

$$\frac{dx}{dt} = ax + by$$
$$\frac{dy}{dt} = cx + dy \qquad \text{where } \begin{vmatrix} a & b \\ c & d \end{vmatrix} \neq 0 \qquad (1)$$

should be stable is that both the characteristic values λ_1, λ_2 should have negative or zero real parts.

Before passing on to nonlinear systems, let us consider a special case of (1). In the second-order equation with constant coefficients

$$\frac{d^2x}{dt^2} + 2k\frac{dx}{dt} + qx = 0 \tag{2}$$

let us as usual take $\dot{x} = dx/dt$ as a new variable; we have then

$$\frac{dx}{dt} = \dot{x}, \qquad \frac{d\dot{x}}{dt} = -qx - 2k\dot{x} \tag{3}$$

This is the special case of (1) when $a = 0$, $b = 1$. We assume $q \neq 0$. If (2) is considered as defining the motion of a dynamical system, the characteristics of (3) represent the trajectories of (2) in the $(x, \dot{x})$ *phase plane*. The first equation of (3) shows that the characteristics of (3) have the following special properties:

(i) A characteristic crossing the $\dot{x}$-axis can do so only in the negative direction, i.e., from quadrant II to quadrant I, or from the quadrant IV to quadrant III. Hence if there exist closed or spiral characteristics, they must circle the origin in the negative direction.

(ii) A characteristic crossing the x-axis must do so at right angles.

We shall briefly enumerate the possible cases; the reader will have no difficulty in verifying our statements.

Case I. $k^2 > q$; λ_1, λ_2 real and distinct. Four rectilinear characteristics, of slopes λ_1 and λ_2.

(a) $q > 0$; $\lambda_1\lambda_2 > 0$. Nodal point; all nonrectilinear characteristics tangent at the origin to the rectilinear characteristic whose slope has the smaller absolute value.

    (1) $k > 0$. Stable equilibrium; rectilinear characteristics in quadrants II and IV.

    (2) $k < 0$. Unstable equilibrium; rectilinear characteristics in quadrants I and III.

(b) $q < 0$; $\lambda_1\lambda_2 < 0$. Saddle point; one rectilinear characteristic in each quadrant.

Case II. $k^2 = q > 0$; λ_1, λ_2 real and equal. Degenerate nodal point. Two rectilinear characteristics in opposite quadrants.

(a) $k > 0$. Stable equilibrium.

(b) $k < 0$. Unstable equilibrium.

Case III. $k^2 < q$; λ_1, λ_2 conjugate complex.

(a) $k = 0$. Vortex point. Ellipses whose axes are the coordinate axes.

(b) $k > 0$. Stable spiral point.

(c) $k < 0$. Unstable spiral point.

The case where λ_1 and λ_2 are complex

$$\frac{d^2x}{dt^2} + 2k\frac{dx}{dt} + qx = 0, \qquad k^2 < q$$

represents damped oscillatory motion. If $k = 0$, the motion is harmonic. If $k > 0$, there is damping, and the amplitude of the oscillations approaches 0. If $k < 0$, the damping is negative and the oscillations become infinite.

As an example of a saddle point, consider the motion of a pendulum about its point of unstable equilibrium, as in Fig. 8. The motion will be given by an equation of the form

$$\ddot{\theta} + 2k\dot{\theta} - q \sin \theta = 0, \qquad q > 0, \qquad k > 0$$

where we are justified (as we shall see in Part C) in substituting θ for $\sin \theta$, as far as discussion of the singularity is concerned. Then $\theta = 0$, $\dot{\theta} = 0$ is a saddle point. If in the position of Fig. 8 we give the pendulum a push of precisely the right magnitude, it will approach equilibrium as $t \to \infty$. This corresponds to one of the trajectories that approaches the origin. It is clear that these trajectories will very seldom be realized in nature.

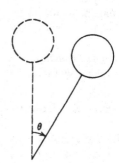

Finally, it should be observed that any second-order equation

$$\frac{d^2x}{dt^2} = F\left(x, \frac{dx}{dt}\right) \tag{4}$$

may be reduced to the form

Fig. 8

$$\frac{dx}{dt} = \dot{x}, \qquad \frac{d\dot{x}}{dt} = F(x, \dot{x}) \tag{5}$$

This is a special case of the general nonlinear system of Part C. If F is defined for all x, $\dot{x}$, the characteristics of (5) will have the properties (i) and (ii) mentioned after equation (3).

PART C.
NONLINEAR SYSTEMS

10. Introduction

We consider now the system

$$\frac{dx}{dt} = P(x, y), \qquad \frac{dy}{dt} = Q(x, y) \tag{1}$$

with an isolated singularity at a point which we may take to be the origin. Throughout this part we consider only singularities of the following type.

Definition 2. *The singularity* $(0, 0)$ *of* (1) *is said to be simple if the following condition is met: There exist constants a, b, c, d, where*

$$\begin{vmatrix} a & b \\ c & d \end{vmatrix} \neq 0$$

such that if $\epsilon_1(x, y)$ *and* $\epsilon_2(x, y)$ *are defined by*

$$
\begin{aligned}
P(x, y) &= ax + by + \epsilon_1(x, y) \\
Q(x, y) &= cx + dy + \epsilon_2(x, y)
\end{aligned}
\tag{2}
$$

then

$$\lim_{\substack{x \to 0 \\ y \to 0}} \frac{\epsilon_i(x, y)}{\sqrt{x^2 + y^2}} = 0, \qquad i = 1, 2 \tag{3}$$

It will appear presently that a simple singularity is isolated. A sufficient condition that $(0, 0)$ be simple is that P and Q should have first partial derivatives continuous in some neighborhood of the origin, with their Jacobian not vanishing at the origin. Then

$$a = \frac{\partial P(0, 0)}{\partial x}, \qquad b = \frac{\partial P(0, 0)}{\partial y}, \qquad c = \frac{\partial Q(0, 0)}{\partial x}, \qquad d = \frac{\partial Q(0, 0)}{\partial y} \tag{4}$$

and (3) is Taylor's theorem.

Now, we know the exact nature of the singularity of

$$\frac{dx}{dt} = ax + by, \qquad \frac{dy}{dt} = cx + dy \tag{5}$$

from our previous discussion. It seems reasonable that the singularity of (1) should have essentially the same nature as that of the system (5) which is derived from it by taking the linear terms of the Taylor expansion for P and Q. We shall show that generally this is in fact the case. We consider the vector fields defined by (1) and (5), adopting as standard notation:

$$
\begin{aligned}
V(x, y) &= (P(x, y),\ Q(x, y)) \\
\bar{V}(x, y) &= (ax + by,\ cx + dy)
\end{aligned}
\tag{6}
$$

Lemma 2. *A simple singularity* $(0, 0)$ *of* V *is isolated. As* $r = \sqrt{x^2 + y^2}$ *approaches* 0 *in any manner we have:*

$$\lim_{r \to 0} \frac{|V(x, y)|}{|\bar{V}(x, y)|} = 1; \qquad \lim_{r \to 0} (\arg V - \arg \bar{V}) = 0 \tag{7}$$

Proof. Since V is continuous and does not vanish on $r = 1$

$$c = \underset{r=1}{\text{g.l.b.}} \; |V| = \underset{r=1}{\text{g.l.b.}} \; [(ax + by)^2 + (cx + dy)^2]^{\frac{1}{2}} > 0$$

Then for all r

$$|V| \geq cr \tag{8}$$

Since by (3)

$$\lim_{r \to 0} \frac{|V - V|}{r} = \lim_{r \to 0} \frac{\sqrt{\epsilon_1^2 + \epsilon_2^2}}{r} = 0 \tag{9}$$

we have

$$\lim_{r \to 0} \left| \frac{V}{|V|} - \frac{V}{|V|} \right| \leq \lim_{r \to 0} \frac{|V - V|}{cr} = 0 \tag{10}$$

From (10) it is clear that there is a neighborhood of the origin in which V cannot vanish, and the formulas (7) follow. We shall frequently use the lemma in this form:

For $\epsilon > 0$ there is a circle $r \leq r_0$ about the origin in which

$$\left| \frac{|V|}{|V|} - 1 \right| < \epsilon, \qquad |\arg V - \arg V| < \epsilon \tag{11}$$

As in Part B, we shall generally perform an affine transformation upon x and y to reduce V to its simplest form. It is clear that this will not affect the validity of any of our work.

For simple singularities we may state the following result regarding characteristics approaching a singularity:

Theorem 2. Let C: $(x(t), y(t))$ represent a characteristic of V approaching the simple singularity $(0, 0)$. A necessary and sufficient condition that C enter $(0, 0)$ with slope m is that

$$\lim_{t \to \infty} \frac{y'(t)}{x'(t)} = m \tag{12}$$

where if $m = \pm\infty$ the limit must properly diverge. The only values m may have are the solutions of

$$\frac{c + dm}{a + bm} = m \tag{13}$$

where $m = \infty$ is a solution if and only if $b = 0$.

Proof. The sufficiency of (12) we have already seen in § 2. Suppose then C enters $(0, 0)$ with slope m

$$\lim_{t \to \infty} \frac{y(t)}{x(t)} = m \tag{14}$$

Then we may write

$$\frac{y'(t)}{x'(t)} = \frac{Q}{P} = \frac{cx + dy + \epsilon_2}{ax + by + \epsilon_1}$$

By (3) we have

$$\lim_{t \to \infty} \frac{y'(t)}{x'(t)} = \lim_{t \to \infty} \frac{c\dfrac{x}{r} + d\dfrac{y}{r}}{a\dfrac{x}{r} + b\dfrac{y}{r}} = \lim_{t \to \infty} \frac{c + d\dfrac{y}{x}}{a + b\dfrac{y}{x}} = \frac{c + dm}{a + bm} \tag{15}$$

from (14). From the existence of the limit in (15) it follows that

$$\lim_{t \to \infty} \frac{y(t)}{x(t)} = \lim_{t \to \infty} \frac{y'(t)}{x'(t)}, \quad \text{i.e., that } m = \frac{c + dm}{a + bm}$$

This completes the proof.

Equation (13) has at most two roots unless $a = d$, $b = c = 0$, when it is satisfied by all real m. Hence just as in the linear case there are at most two directions in which a characteristic can enter the origin, except for the completely degenerate nodal point of § 6. Theorem 2 of course holds for linear systems as a special case.

Throughout the next three sections we shall adopt the following notation as standard. $\phi(x, y)$ will denote the angle (with the proper sign) between the vector $V(x, y)$ to a point $A: (x, y)$ and the radius vector $\overline{AO}$; see Fig. 9. We make the definition unique by specifying

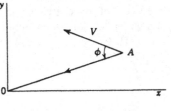

Fig. 9

$$-\pi < \phi \leq \pi$$

$\psi(x, y)$ will denote the corresponding angle for $\overline{V}$. Observe that $\psi(kx, ky) = \psi(x, y)$, $k \neq 0$. For any $\epsilon > 0$ there exists a circle $r \leq r_0$ in which

$$|\phi - \psi| < \epsilon$$

It will frequently be convenient to represent the characteristics of V in *polar form*

$$r = r(t), \qquad \theta = \theta(t)$$

where $r(t)$ and $\theta(t)$ will then satisfy the equations

$$\frac{dr}{dt} = -|V| \cos \phi, \qquad \frac{d\theta}{dt} = \frac{|V| \sin \phi}{r} \tag{16}$$

11. Nodal Points

We consider first the case where V has a nodal point, i.e., where the roots of

$$\begin{vmatrix} a - \lambda & b \\ c & d - \lambda \end{vmatrix} = 0$$

are real, unequal, and of the same sign; say $\lambda_1 < \lambda_2 < 0$. Then by a suitable affine transformation we may take the vector fields in the canonical form:

$$V = (\lambda_1 x + \epsilon_1(x, y), \lambda_2 y + \epsilon_2(x, y))$$

$$\bar{V} = (\lambda_1 x, \lambda_2 y)$$

(1)

where

$$\lim_{r \to 0} \frac{\epsilon_i}{r} = 0$$

From our conditions on λ_1 and λ_2 it follows that everywhere $|\psi| < \pi/2$ (see Fig. 2 of this chapter), and since ψ is continuous and a function only of y/x, $|\psi|$ has an upper bound $\psi_0 < \pi/2$. Also, in the open quadrants I and III, $0 < \psi \leq \psi_0 < \pi/2$ and in II and IV, $-\pi/2 < -\psi_0 \leq \psi < 0$. See Fig. 10.

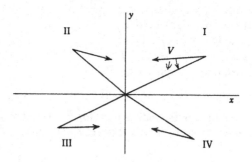

Fig. 10

For any $\epsilon > 0$, there is a circle Γ: $r \leq r_1$ in which V vanishes nowhere and $|\phi - \psi| < \epsilon$. Take $\epsilon = \frac{1}{2}(\frac{1}{2}\pi - \psi_0)$. Then in Γ

$$|\phi| < \psi_0 + \epsilon = \frac{\pi}{2} - \epsilon$$

(2)

Let C: $(r(t), \theta(t))$, $t \geq t_0$, be any positive half-characteristic of V starting in $r \leq r_1$ expressed in polar form. Then

$$\frac{dr}{dt} = -|V| \cos \phi, \qquad \frac{d\theta}{dt} = \frac{|V| \sin \theta}{r} \qquad (3)$$

Since by (2) $\cos \phi > 0$ in Γ, $r(t)$ is monotonically decreasing. Therefore C cannot get out of Γ, and hence by Lemma 1 is defined for all $t \geq t_0$. Since this is also true for any circle $r \leq r_2 < r_1$, the equilibrium is certainly stable. Suppose C does not approach the origin. Then, since $r(t)$ is monotonic and does not approach 0, it has a g.l.b. $r_0 > 0$.

$$r_0 < r(t) \leq r_1, \qquad t \geq t_0 \qquad (4)$$

But in $r_0 \leq r \leq r_1$, $|V|$ has a lower bound $k > 0$; therefore by (3)

$$\frac{dr}{dt} \leq -k \cos \left(\frac{\pi}{2} - \epsilon \right)$$

$$r \leq -k \cos \left(\frac{\pi}{2} - \epsilon \right)(t - t_0) + r_1, \qquad t \geq t_0 \qquad (5)$$

Then for t sufficiently large, $r(t) < r_0$, contrary to our assumption. Therefore every half-characteristic starting in Γ remains in Γ and approaches the origin.

By Theorem 2 we see that a characteristic of V can enter the origin only along one of the co-ordinate axes. We shall prove first that every characteristic starting in Γ enters the origin; precisely: *If a positive half-characteristic of V starting in Γ does not enter the origin along the x-axis, it enters along the y-axis.*

Let δ be any positive number. Let us surround the x- and y-axes with sectors of angle δ, and label the regions of the plane so defined as in Fig. 11.

Now it is easily verified that there is a $\psi_1 > 0$ which depends on δ such that in all the B-regions, $|\psi| \geq \psi_1(\delta) > 0$. Then for every $\delta > 0$, there exists a circle $r \leq \rho(\delta)$ in which $|\phi - \psi| < \frac{1}{2}\psi_1(\delta)$. Let us define $A_1(\delta)$, $B_1(\delta)$, etc., as these parts of A_1, B_1, etc., lying in $r \leq \rho(\delta)$. Then in $B_i(\delta)$, $|\phi| \geq \frac{1}{2}\psi_1(\delta) > 0$; where in $B_1(\delta)$ and $B_3(\delta)$, $\phi > 0$, and in $B_2(\delta)$ and $B_4(\delta)$, $\phi < 0$. Therefore on the rectilinear boundaries of $A_1(\delta)$ and $A_3(\delta)$, V points *into* a B-region; and on the rectilinear boundaries of $A_2(\delta)$ and $A_4(\delta)$, V points *out* of a B-region. Since no half-characteristic starting in $r \leq \rho(\delta)$ can get out of it, we can make the following statements:

(a) No characteristic starting in $A_2(\delta)$ or $A_4(\delta)$ can get out of it.

(b) No characteristic starting in $r \leq \rho(\delta)$, but not in $A_1(\delta)$, can get into $A_1(\delta)$; and likewise for $A_3(\delta)$.

Let now C: $(x(t), y(t))$ be a positive half-characteristic starting in $r \leq r_1$. Then we observe: a necessary and sufficient condition that C enter the origin along the x-axis on the right is that for every $\delta > 0$ there exists a number $t(\delta)$ such that for $t \geq t(\delta)$, C lies entirely in $A_1(\delta)$. *A similar remark holds for the other three possible methods of entrance.*

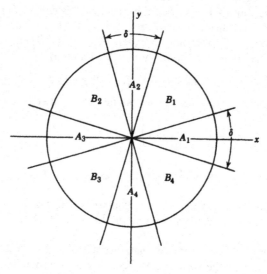

Fig. 11

Suppose now that C is a half-characteristic starting in $r \leq r_1$ which *does not enter the origin along the x-axis.* Then by the remark above, for some value δ_0, C lies in one of the regions $B_i(\delta_0)$; say without loss of generality $B_1(\delta_0)$. Now assume that for *every* $\delta \leq \delta_0$, C *eventually enters* $A_2(\delta)$. Suppose this were not so for some value of δ, say δ_1. Then for all values of $t \geq t_1$, say, C remains in $B_1(\delta_1)$. But in $B_1(\delta_1)$, by (3)

$$\frac{d\theta}{dt} = \frac{|V| \sin \phi}{r} > 0$$

since by construction ϕ was positive in $B_1(\delta_1)$. Therefore $\theta(t)$ is monotonic, and since $\theta(t) \leq \frac{1}{2}(\pi - \delta_1)$

$$\lim_{t \to \infty} \theta(t) = \theta_0 < \frac{\pi}{2} \qquad (6)$$

This implies that C enters the origin along the line of slope $\tan \theta_0$, whereas we know that C can enter the origin only along one of the co-ordinate axes. Therefore C enters every $A_2(\delta)$; and since it must stay in every $A_2(\delta)$, it enters the origin along the y-axis from above. Similar remarks hold if C is ever in one of the other B-regions.

We have thus proved the following statement: Every half-characteristic starting in Γ: $r \leq r_1$ enters the origin, and every half-characteristic that for any value of t is in one of the regions $B_1(\delta)$ or $B_2(\delta)[B_3(\delta)$ or $B_4(\delta)]$ enters the origin along the y-axis from above [from below].

We prove now (by an elementary topological argument) that at least one characteristic enters the origin along the x-axis on either side. Suppose, on the contrary, that there is no half-characteristic entering the origin along the x-axis from the right. Then consider the set $R = A_1(\delta_0)$, any δ_0, the origin not included. R is known to be connected. Let M be the set of all points (x, y) of R such that the characteristic passing through (x, y) enters the origin along the y-axis from above, and N the set of points of R such that that characteristic enters the origin along the y-axis from below. Then

$$M + N = R; \qquad M \cdot N = 0; \qquad M, N \neq 0 \tag{7}$$

Suppose now (x_0, y_0) lies in R. Let C: $(x(t), y(t))$ be the characteristic passing through (x_0, y_0) for $t = 0$; then for $t = t_1$, say, the point $(x(t_1), y(t_1))$ lies in the interior of some $B_i(\delta)$, for some δ. Then by the continuity of characteristics in their starting point, there exists some neighborhood N_0 of (x_0, y_0) such that every characteristic starting in N_0 enters $B_i(\delta)$ and hence enters the origin along the y-axis in the same direction as C. That is to say, M and N are *open in R*. But this, together with (7), implies that R is not connected, a contradiction. Hence our assertion.

If now two characteristics C_1 and C_2 enter the origin along the x-axis on the right, clearly any characteristic starting between them within $r \leq r_1$ also enters along the x-axis. Also, if we denote now by M the set of all points (x, y) within $r \leq r_1$ such that the characteristic through (x, y) enters the origin along the y-axis, by the preceding argument M is open in R; wherefore its complement in R is closed. That is to say, of all characteristics that enter the origin along the x-axis, there is a "last" one on either side, in the following sense: if we consider the points of the circumference of $r \leq r_1$, those through which the characteristic enters the origin along the x-axis are distributed in two closed arcs, as shown in Fig. 12.

Thus we are certainly justified in calling $(0, 0)$ a *nodal point*. If sufficient further restrictions are put upon P and Q, it is possible to prove that there exists *precisely one* characteristic entering the origin along the x-axis on either side. The characteristics will then present precisely the same configuration as for a linear system (Fig. 13). The "normal" direction of approach (in our previous case, the y-axis) corresponds to the root λ of the characteristic equation with smaller absolute value.

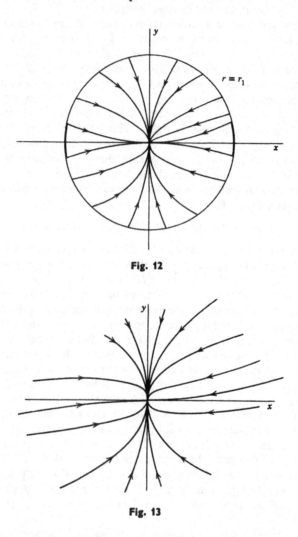

Fig. 12

Fig. 13

In the general case, the pictures will be affinely distorted but the configurations will remain unchanged. In the course of our work we proved that if $r = r_1$ is a sufficiently small circle, then every characteristic intersecting it must cross from its outside to inside. In the general case, these circles will become ellipses with their axes in certain directions. They are the curves without contact of which we shall speak in the next chapter.

Finally, if $\lambda_1 > \lambda_2 > 0$, the singularity will be an unstable nodal point, and as $t \to -\infty$ we have the same picture as before.

12. Saddle Points

Suppose now the linear vector field $\bar{V}$ has a saddle point at $(0, 0)$; then the characteristic values will be real and of opposite sign, say

$$\lambda_1 < 0 < \lambda_2$$

We may then assume the canonical form

$$\bar{V} = (\lambda_1 x, \lambda_2 y)$$
$$V = (\lambda_1 x + \cdot \epsilon_1, \lambda_2 y + \epsilon_2) \tag{1}$$

The only directions along which a characteristic can enter the origin are the co-ordinate axes.

The angle ψ between $\bar{V}$ and the radius vector $\overline{AO}$ (Fig. 14) will behave as follows:

1. In quadrants I and III, $\psi > 0$.
2. In quadrants II and IV, $\psi < 0$.
3. On the x-axis, $\psi = 0$.
4. On the y-axis, $\psi = \pi$.

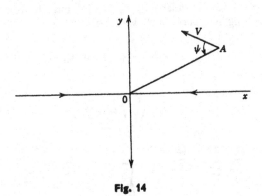

Fig. 14

It is clearly possible to surround the y-axis by a sector S of angle $\gamma > 0$, such that in S, $|\psi| \geq \pi/2 + 2\epsilon$, $\epsilon > 0$. Then in some circle $r \leq r_1$, V

does not vanish (except at the origin) and $|\phi - \psi| < \epsilon$. Let S_1 be the part of S in $r \leq r_1$; then in S_1 (Fig. 15), $|\phi| > \pi/2 + \epsilon$.

We prove that the origin represents an *unstable* equilibrium. Let C: $(r(t), \theta(t))$, $t \geq t_0$, be any half-characteristic starting in S_1, where $r(t_0) = r_0$,

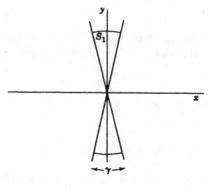

Fig. 15

say. Then C can only get out of S_1 by going out of $r = r_1$; for on the rectilinear boundaries of S_1, V points *into* S_1. But in S_1 we have

$$\frac{dr}{dt} = -|V| \cos \phi \geq -|V| \cos\left(\frac{\pi}{2} + \epsilon\right) \tag{2}$$

and in that part of S_1 for which $r_0 \leq r \leq r_1$, $|V|$ has a lower bound $k > 0$. Therefore as long as C stays in S_1, we have

$$r(t) \geq -k \cos\left(\frac{\pi}{2} + \epsilon\right)(t - t_0) + r_0 \tag{3}$$

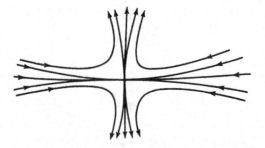

Fig. 16

and C must eventually go out of S_1. Therefore the equilibrium cannot be stable.

It can be shown that any characteristic not entering the origin cannot approach it, and that any characteristic not approaching the origin must

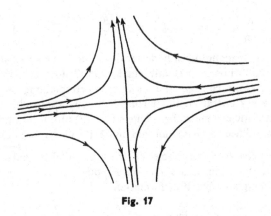

Fig. 17

go out of the circle $r \leq r_1$. We may also show (as in § 11) that at least one characteristic must enter the origin along the x-axis on either side, and that all such characteristics form a continuum. Similarly, there will be at least two characteristics approaching the origin along the y-axis for $t = -\infty$. The singularity is therefore a saddle point, and we have a picture of the form shown in Fig. 16. Under further conditions on $P(x, y)$ and $Q(x, y)$ it can be shown that precisely two characteristics enter for $t = -\infty$, with that of the linear case illustrated in Fig. 17.

13. Spiral Points

Suppose finally the linear system V has a spiral point; i.e., the characteristic values λ_1, λ_2 are conjugate complex: say $\lambda = \alpha \pm i\beta$, α, $\beta \neq 0$. For definiteness take $\alpha < 0$. We may then take the vector fields in the canonical form

$$V = (\alpha x - \beta y + \epsilon_1, \ \beta x + \alpha y + \epsilon_2)$$
$$\bar{V} = (\alpha x - \beta y, \ \beta x + \alpha y) \tag{1}$$

The characteristics of $\bar{V}$ are a family of logarithmic spirals approaching the origin as $t \to \infty$. It is known that every radius vector will cut a logarithmic spiral in the same angle ψ, and since all the spirals are similar ψ will be a constant, $|\psi| < \pi/2$.

Take $\epsilon = \frac{1}{2}(\frac{1}{2}\pi - |\psi|)$. Then for some circle $r \leq r_0$

$$|\phi - \psi| < \epsilon, \qquad |\psi| - \epsilon < |\phi| < |\psi| + \epsilon = \frac{\pi}{2} - \epsilon$$

Let C: $(r(t), \theta(t))$, $t \geq t_0$, be any half-characteristic starting in $r \leq r_0$. Then

$$\frac{dr}{dt} = -|V| \cos \phi < 0; \qquad \frac{d\theta}{dt} = \frac{|V| \sin \phi}{r} > 0 \qquad (2)$$

Therefore $r(t)$ is monotonically decreasing, and C stays in $r \leq r_0$, wherefore the equilibrium is certainly stable. If C does not approach the origin, we prove a contradiction as in the corresponding proof of § 11. Therefore every characteristic starting in $r \leq r_0$ approaches the origin. No characteristic can enter the origin, since (13) of § 10 has no real roots in our case. Since $\theta(t)$ is monotonic, by (2) it follows that $\lim_{t \to \infty} \theta(t) = \infty$, i.e., that the characteristics circle the origin an infinite number of times.

Therefore the origin is a stable spiral point. If we write the system without making an affine transformation

$$\frac{dx}{dt} = ax + by + \epsilon_1, \qquad \frac{dy}{dt} = cx + dy + \epsilon_2 \qquad (3)$$

the direction of rotation of the spirals will be the same as the sign of c. If $\alpha = \frac{1}{2}(a + d) > 0$, the same analysis will hold except that the equilibrium is unstable.

Observe that here again the circles $r = r_1 \leq r_0$ are curves without contact; for the general system (3) they become a certain family of ellipses.

14. Indeterminate Cases

The cases where the linear system V has a degenerate nodal point are of minor importance, and we omit them. However, if V has a vortex point, the behavior of the nonlinear system is rather interesting. We may reduce to canonical form

$$\frac{dx}{dt} = -\beta y + \epsilon_1(x, y), \qquad \frac{dy}{dt} = \beta x + \epsilon_2(x, y) \qquad (1)$$

In polar form

$$\frac{dr}{dt} = \frac{1}{r}\left(x \frac{dx}{dt} + y \frac{dy}{dt}\right) = \frac{\epsilon_1 x + \epsilon_2 y}{r}$$

$$\frac{d\theta}{dt} = \frac{1}{r^2}\left(x \frac{dy}{dt} - y \frac{dx}{dt}\right) = \beta + \frac{\epsilon_2 x - \epsilon_1 y}{r^2} \qquad (2)$$

Therefore for r small, dr/dt is very near 0 and $d\theta/dt$ is very near β. The nature of the singularity is indeterminate since we do not know the sign of dr/dt as in the previous sections. There are a number of possible cases:

Case I. $(0, 0)$ may be a vortex point. This is true when (1) is linear. Consider also the equation

$$\frac{d^2y}{dt^2} + \left(\frac{dy}{dt}\right)^2 + y = 0 \tag{3}$$

We may write (3) in the form

$$\frac{dx}{dt} = -y - x^2, \qquad \frac{dy}{dt} = x \tag{4}$$

This is of type (1). Suppose in $r < r_0$ that $d\theta/dt > \frac{1}{2}$. Consider the characteristic C of (4) passing through the point $(x_1, 0)$, $x_1 > 0$, for $t = 0$. Then for x_1 sufficiently small, C will remain in $r < r_0$ for $|t| \leq 2\pi$ by virtue of (2) with $\epsilon_1 = -x^2$ and $\epsilon_2 = 0$. Therefore C will intersect the negative x-axis at some point $(x_2, 0)$. But since (4) is unchanged if we replace t by $-t$ and y by $-y$, it follows that the *negative* characteristic starting from $(x_1, 0)$ also passes through $(x_2, 0)$, i.e., that C is closed. Since this is true for all characteristics sufficiently close to $(0, 0)$, the singularity is a vortex point.

Case II. Consider now the equation

$$\frac{d^2y}{dt^2} + \left(\frac{dy}{dt}\right)^3 + y = 0$$

i.e., the system

$$\frac{dx}{dt} = -y - x^3, \qquad \frac{dy}{dt} = x \tag{5}$$

Here

$$\frac{dr}{dt} = -\frac{x^4}{r}$$

Therefore except on the y-axis, $dr/dt < 0$, and it is easy to see that no characteristic can be closed. By a familiar argument, it follows that every characteristic must approach the origin; and since $d\theta/dt$ is positive and bounded away from 0 in some circle about $(0, 0)$, the origin must be a stable spiral point.

In the same way, the origin is an unstable spiral point of

$$\frac{d^2y}{dt^2} - \left(\frac{dy}{dt}\right)^3 + y = 0$$

Case III. Consider finally the system

$$\frac{dx}{dt} = -y + xr^2 \sin \frac{\pi}{r} = P(x, y)$$

$$\frac{dy}{dt} = x + yr^2 \sin \frac{\pi}{r} = Q(x, y)$$

(6)

This has no singularity but $(0, 0)$. It is readily verified that P and Q have continuous first derivatives everywhere, including the origin. In polar form (6) becomes

$$\frac{dr}{dt} = r^3 \sin \frac{\pi}{r}, \qquad \frac{d\theta}{dt} = 1$$

(7)

The circles $r = 1/n$ $(n = 1, 2, \cdots)$ are closed characteristics. Furthermore

$$\frac{dr}{dt} > 0 \qquad \qquad r > 1$$

$$\frac{dr}{dt} < 0 \qquad \frac{1}{2m} < r < \frac{1}{2m - 1} (m = 1, 2, \cdots)$$

(8)

$$\frac{dr}{dt} > 0 \qquad \frac{1}{2m + 1} < r < \frac{1}{2m}$$

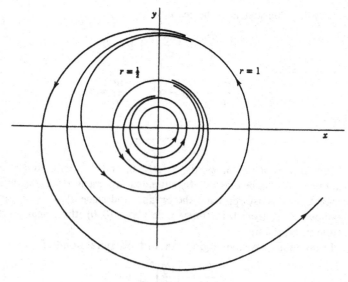

Fig. 18

Therefore no characteristic but the circles $r = 1/n$ can be closed; for any characteristic must remain in one of the regions $r > 1$, $1/2m < r < 1/(2m - 1)$, $1/(2m + 1) < r < 1/2m$ $(m = 1, 2, \cdots)$; and in each of these regions r is strictly monotonic. Furthermore every nonclosed characteristic must either approach one of the circles as $t \to \pm\infty$ or become unbounded; for in every closed bounded region R of the plane not containing any point for which $r = 1/n$, $|dr/dt|$ is bounded away from zero, and a characteristic starting in R must go out of R. Finally, nonclosed characteristics must approach the circles spirally, since $d\theta/dt = 1$. The circles $r = 1/n$ are stable or unstable accordingly as n is even or odd. See Fig. 18.

Such a singularity, with both closed and nonclosed characteristics arbitrarily close to it, is known as a *center*. By ingenuity, very complicated centers may be constructed.

5

Solutions of
an Autonomous System
in the Large

1. Introduction

In this chapter we consider the geometrical picture presented by the characteristics of

$$\frac{dx}{dt} = P(x, y), \qquad \frac{dy}{dt} = Q(x, y)$$

over the entire domain D in which P and Q are defined. We shall assume an elementary knowledge of the geometry of the plane, including the Jordan curve theorem. Our main result will be the classical theorem of Poincaré-Bendixson on the existence of closed characteristics ("limit cycles"). To avoid confusion, we shall use the term *cyclic characteristic* to describe a closed characteristic curve representing a periodic solution, and use *closed* to denote topological closure only.

Consider, for example, the system

$$\frac{dx}{dt} = y + x(1 - x^2 - y^2), \qquad \frac{dy}{dt} = -x + y(1 - x^2 - y^2) \qquad (1)$$

Transforming to polar co-ordinates we get

$$\frac{dr}{dt} = r(1 - r^2), \qquad \frac{d\theta}{dt} = -1 \tag{2}$$

It is easily verified that the most general solution of (2) is

$$r = \frac{1}{\sqrt{1 + ke^{-2t}}}, \qquad \theta = -(t - t_0) \tag{3}$$

with k and t_0 arbitrary. Let us set $t_0 = 0$ for convenience; then every solution of (1) may be written in the form

$$x = \frac{\cos t}{\sqrt{1 + ke^{-2t}}}, \qquad y = \frac{-\sin t}{\sqrt{1 + ke^{-2t}}} \tag{4}$$

If $k = 0$, (4) is the circle $x^2 + y^2 = 1$. If $k = +c^2$, (4) represents a spiral approaching the origin as $t \to -\infty$, and approaching the circle spirally from the inside as $t \to +\infty$. If $k = -c^2$, (4) represents a curve approaching ∞ as $t \to +\log c + 0$, and approaching the circle from the outside as $t \to +\infty$. We have thus a picture of the form shown in Fig. 1.

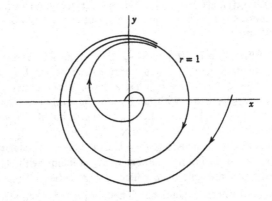

Fig. 1.

Therefore there exists a single cyclic characteristic, which all the non-cyclic characteristics approach spirally as $t \to +\infty$. Such a cyclic characteristic is known as a *limit cycle* after Poincaré. By the methods of this chapter, the existence of limit cycles may be proved for a general class of equations of physical interest.

2. Geometrical Considerations

Let $P(x, y)$ and $Q(x, y)$ have continuous first-order partial derivatives in a certain domain D, and let $V(x, y) = (P, Q)$ be the vector field they define.

Definition 1. A segment without contact (with respect to V) is a finite closed segment L of a straight line, such that
(a) *Every point of L is a regular point of V.*
(b) *At no point of L does the vector V have the same direction as L.*

That is, a segment without contact passes through no singularity of V and is tangent to no characteristic of V. We proceed to develop some elementary properties of such segments.

(i) Through any regular point A: (x_0, y_0) of D there may be drawn a segment L without contact containing A in its interior, and L may have any direction except that of $V(x_0, y_0)$.

(ii) Every characteristic intersecting a s.w.c. L actually crosses from one side of L to the other, and all characteristics crossing L do so in the same direction.

The proofs of (i) and (ii) follow immediately from the continuity of V and the uniqueness of characteristics.

(iii) If C: $(x(t), y(t))$, $a \leq t \leq b$, is a finite arc of a characteristic of V, then C cannot cross a given segment without contact L more than a finite number of times.

Proof of (iii). For let $\{t_n\}$ $(n = 1, 2, \cdots)$ be an infinite sequence of different values of t, $a \leq t_n \leq b$, representing points $\{A_n\}$ of the intersection of L and C. Then $\{t_n\}$ has a subsequence converging to some number t_0, $a \leq t_0 \leq b$; without loss of generality we may take this subsequence to be $\{t_n\}$ itself. Let A_0 be the point of C given by $t = t_0$; clearly A_0 is on L. Then the sequence $\{A_n\}$ approaches A_0. If more than one point A_n is the same as A_0, then C is a cyclic characteristic of period $h > 0$. Hence if an infinite number of A_n coincide with A_0, then $b - a$ must exceed any multiple of h, which is impossible. Therefore we may take every A_n different from A_0. Then $\overline{\{A_0 A_n\}}$ is a sequence of secants whose limiting direction is the tangent to C at A_0. But this limiting direction is the direction of L; wherefore L is tangent to C at A_0, a contradiction. Hence there can be at most a finite number of intersections of C and L.

(iv) Let A be an interior point of a segment L without contact. Then for every $\epsilon > 0$ there exists a circle Γ with center A, such that every characteristic passing through a point of Γ for $t = 0$ intersects L for $t = $ some number t_0, where $|t_0| < \epsilon$.

Proof of (iv). Let A be the origin of (x, y) co-ordinates and L be on the x-axis. The solutions $(x(t, x_0, y_0), y(t, x_0, y_0))$ of

$$\frac{dx}{dt} = P(x, y), \qquad \frac{dy}{dt} = Q(x, y)$$

have continuous first derivatives with respect to t, x_0, y_0 in some sphere $t^2 + x_0^2 + y_0^2 < \rho^2$. Moreover $[\partial y(0, 0, 0)]/\partial t \neq 0$ by the definition of a segment without contact. Therefore by the implicit function theorem the equation $y(t, x_0, y_0) = 0$ has a unique continuous solution $t = t(x_0, y_0)$ for (x_0, y_0) in some circle Γ about $(0, 0)$. Since $t(0, 0) = 0$, for Γ sufficiently small we shall have $|t(x_0, y_0)| < \epsilon$ for (x_0, y_0) in Γ, and the proof is complete.

Suppose now that C: $(x(t), y(t))$, $t \geq t_0$, is a positive half-characteristic of V, lying in a closed bounded region $R \subset D$. Then by Lemma 1, Chapter 4, C is defined for all $t \geq t_0$.

We shall say that a point (x_0, y_0) is a *limit point* of C if there exists a sequence $\{t_n\} \to \infty$ such that the sequence of points $\{x(t_n), y(t_n)\}$ converges to (x_0, y_0). C must have at least one limit point; for the sequence of points $\{x(t_0 + n), y(t_0 + n)\}$ $(n = 1, 2, \cdots)$, being contained in the closed bounded region R, has a convergent subsequence. The set of all limit points of C is denoted by C'; clearly $C' \subset R$.

3. Basic Lemmas

We assume throughout this section and the next that C: $(x(t), y(t))$, $t \geq t_0$, is a positive half-characteristic contained in a closed bounded region $R \subset D$.

Lemma 1. The limit set C' of C is closed and connected; and if $d(M; N)$ is the distance between the point sets M and N, we have

$$\lim_{t \to \infty} d[(x(t), y(t)); C'] = 0 \tag{1}$$

Proof. Suppose $\{A_n\}$ is a sequence of points of C' converging to the point A. Then there exists a sequence $\{t_n\}$, where $t_n > n$, and

$$d[(x(t_n), y(t_n)), A_n] < \frac{1}{n}$$

Then the sequence $\{x(t_n), y(t_n)\}$ converges to A, whence A is a point of C'; wherefore C' is closed.

Suppose C' is not connected. Then we can represent it as a sum of two disjoint nonnull sets M and N, each closed in C'

$$C' = M + N; \qquad M \cdot N = 0; \qquad M, N \neq 0 \tag{2}$$

Since C' is closed M and N are closed, and since M and N are disjoint, closed, and bounded, they have a nonzero distance δ. Now there exists a monotonic sequence $\{t_n\} \to \infty$ such that

$$\text{with } n \text{ odd,} \qquad d[\{x(t_n), y(t_n)\}; \; M] < \frac{\delta}{4}$$

$$\text{with } n \text{ even,} \qquad d[\{x(t_n), y(t_n)\}; \; N] < \frac{\delta}{4} \tag{3}$$

Therefore by the continuity of $x(t)$ and $y(t)$ there exists a sequence $\{t_n'\}$ such that

$$t_{2n-1} < t_n' < t_{2n}; \qquad n = 1, 2, \cdots, d[\{x(t_n'), y(t_n')\}; \; M]$$
$$= d[\{x(t_n'), y(t_n')\}; \; N]$$

hence

$$d[\{x(t_n'), y(t_n')\}; \; \{M + N\}] \geq \frac{\delta}{2} \tag{4}$$

The sequence of points $\{x(t_n'), y(t_n')\}$ must have a subsequence converging to a limit point $(\bar{x}, \bar{y})$ of C'; but by (4)

$$d[(\bar{x}, \bar{y}); \; M + N] \geq \frac{\delta}{2} \tag{5}$$

a contradiction. Therefore C' must be connected.

Finally, if (1) does not hold, there must exist a number $\delta > 0$ and a sequence $\{t_n\} \to \infty$ such that

$$d[\{x(t_n), y(t_n)\}; \; C'] \geq \delta$$

Then as from (4) above, a contradiction is easily found; thus all parts of the lemma have been proved.

Lemma 2. *If C' contains a regular point A: $(\bar{x}, \bar{y})$, the characteristic Γ through A lies entirely in C'.*

Proof. Let $(\bar{x}(t), \bar{y}(t))$ represent Γ, where

$$(\bar{x}(0), \bar{y}(0)) = (\bar{x}, \bar{y}) \tag{6}$$

Let $\{t_n\} \to \infty$ be such a sequence that the sequence of points

$$\{x(t_n), y(t_n)\} \to (\bar{x}, \bar{y}) \tag{7}$$

Define new representations of C by

$$(x_n(t), y_n(t)) = (x(t + t_n), y(t + t_n)) \tag{8}$$

Then from (6) and (7) we have

$$(x_n(0), y_n(0)) \to (\bar{x}(0), \bar{y}(0)) \tag{9}$$

Now let $(\bar{x}(t_0), \bar{y}(t_0))$ be any point of Γ. It follows from (9) by the continuity of solutions in their initial values that

$$(x(t_0 + t_n), y(t_0 + t_n)) = (x_n(t_0), y_n(t_0)) \to (\bar{x}(t_0), \bar{y}(t_0)) \tag{10}$$

Since $t_0 + t_n \to \infty$, $(\bar{x}(t_0), \bar{y}(t_0))$ is in C' and the lemma follows.

Therefore C' consists of a set of singularities and a set of complete characteristics. Observe that as yet we have not used the topology of the plane in any way; our results in fact hold for a vector field in n-dimensional space. The following lemma, however, depending as it does on Jordan's curve theorem, is peculiar to the plane.

Lemma 3. *Suppose C' contains a regular point A. Then if L is a segment without contact through A, there exists a monotonic sequence $\{t_n\} \to \infty$ such that the points of intersection of C and L are precisely $\{A_n\} = \{x(t_n), y(t_n)\}$. If $A_1 = A_2$, then $A = A_n$ $(n = 1, 2, \cdots)$ and C is cyclic. If $A_1 \neq A_2$, all the A_n's are different and on L, and A_{n+1} is between A_n and A_{n+2}, $(n = 1, 2, \cdots)$.*

Proof. By § 2(i) there exists a segment L without contact passing through A. Now every circle about A contains points of C corresponding to arbitrarily large values of t. Therefore by § 2(iv) L is cut by C in points corresponding to arbitrarily large values of t; i.e., there are an infinite number of values of t corresponding to points of intersection of C and L. But by § 2(iii) any finite arc of C can cut L in only a finite number of points. Therefore the points of intersection may be represented by an infinite monotonic sequence $\{t_n\} \to \infty$ as required. Let $\{A_n\}$ be defined as in the statement of the theorem. If $A_1 = A_2$, C is cyclic, $A_1 = A_n$, and since A is an accumulation point of the A_n's, $A = A_1 = A_n$. It remains, therefore, to prove the last statement; and only in this part of the lemma is the topology of the plane essential to the proof.

Suppose then $A_1 \neq A_2$. C does not intersect L for $t_1 < t < t_2$; therefore that part of C given by $t_1 \leq t \leq t_2$, plus the segment $\overline{A_1 A_2}$, forms a simple closed Jordan curve Γ. There are two cases:

Case I. For $t > t_2$, C is *inside* Γ as shown in Fig. 2. Then to get outside Γ, C must cross Γ, by Jordan's theorem; but C cannot cross itself, and it cannot cross $\overline{A_1 A_2}$ in the wrong direction [cf. § 2(ii)]. Therefore C remains inside Γ for all $t > t_2$. Hence it is clear that A_3 is different from A_1 and A_2, and that A_2 is between A_1 and A_3. The statement of the lemma follows by induction.

Case II. For $t > t_2$, C is *outside* Γ as shown in Fig. 3. Then C cannot again get inside Γ, and the argument proceeds as before with obvious modifications.

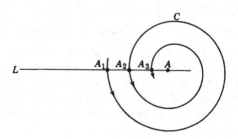

Fig. 2

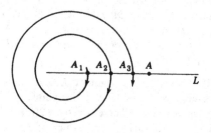

Fig. 3

It follows that the sequence of points $\{A_n\}$ has only the one accumulation point A. From this remark we have immediately:

Lemma 4. *No segment L without contact can cut C' in two distinct points.*

Lemma 5. *If C' contains a cyclic characteristic F, it contains no other point.*

Proof. Suppose on the contrary that for some C, C' contains a cyclic characteristic F and at least one other point. $C' - F$ is not null, and hence cannot be closed; for otherwise C' would be the sum of two closed disjoint sets ($C' - F$ and F) and hence not connected, contrary to Lemma 1. Since C' is closed, there exists a point A on F which is an accumulation point of points of $C' - F$. Let L be a segment without contact through A. Since every circle about A contains a point of $C' - F$, by § 2(iv) it is clear that L must be cut by a characteristic belonging to $C' - F$. Therefore L cuts C' in two distinct points, contrary to Lemma 4. This completes the proof.

4. Theorem of Poincaré-Bendixson

We now have all the materials to prove our main results. We assume as before that C: $(x(t),\ y(t))$, $t \geq t_0$, is a positive half-characteristic contained in a closed bounded region $R \subset D$.

Theorem 1. (*Poincaré-Bendixson*). *If C' contains no singularity of V, then either:*

(1) $C (= C')$ *is a cyclic characteristic, or*

(2) C' *consists of a cyclic characteristic F (a "limit cycle") that C approaches spirally from either the inside or outside.*

Proof. Let A be a point of C', and F the characteristic through A. Let F' be the limit set of F. Since C' is closed, $F' \subset C'$. F' contains at least one point B which (being in C') must be regular. Let L be a segment without contact through B. By Lemma 4, L cuts F in precisely one point (namely B). Therefore by Lemma 3, F is cyclic. By Lemma 5, $C' = F$, wherefore in any case C' consists of the single cyclic characteristic F.

If C is cyclic, $C = F$ and we have Case I of the theorem. If C is not cyclic, C is either inside or outside F. Let A be a point of F, L a segment without contact through A, and $\{A_n\}$ ($n = 1, 2, \cdots$) the sequence of points of intersection of L and C, corresponding to the monotonic sequence $\{t_n\} \to \infty$ (see Lemma 3). Then by simple arguments based on Jordan's theorem, we see that according as C is inside or outside F we have the configurations shown in Fig. 4. Therefore the approach is certainly spiral.

We remark further that by considerations of continuity it is not too hard to show that

$$\lim_{n \to \infty} (t_{n+1} - t_n) = h \tag{1}$$

where h is the period of F. Thus the motion on C as $t \to \infty$ differs arbitrarily little from the periodic motion represented by F. Finally, if we replace t by $-t$, we may obtain analogous results as to the behavior of characteristics when $t \to -\infty$.

In the course of our work we showed that if C' contains a point of C, C is necessarily cyclic. This is not true for higher dimensions and the analogue of the Poincaré-Bendixson theorem does not hold.

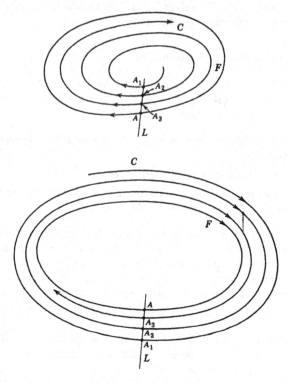

Fig. 4

Theorem 2. *Suppose the singularities of V are isolated and C' contains a singularity A. Then*

(1) *If C' contains no regular point, C' consists solely of the point A, which C approaches as $t \to \infty$.*

(2) *If C' contains a regular point, C' consists of a finite set of singularities $\{A_n\}$ and a set of characteristics $\{C_a\}$, where as $t \to \pm\infty$ each C_a approaches a singularity.*

Proof. If C' has no regular point, being connected it can contain only A, and by Lemma 1, C approaches A as $t \to \infty$. If C' contains a regular point, C' consists of a set $\{A_n\}$ of singularities and a set $\{C_a\}$ of characteristics, where $\{A_n\}$ must be finite since the region R in which C' is contained is bounded and closed. If any C_a' contains a regular point, C_a must be cyclic as in the proof of Theorem 1; which by Lemma 5 contradicts the assumption that C' contains a singularity. Therefore no C_a' contains a regular point; hence every C_a approaches a singularity as $t \to +\infty$, and likewise as $t \to -\infty$.

Suppose now that there exist two concentric closed curves C_1 and C_2, bounding an annulus A free from singularities, and that on C_1 and C_2 V everywhere points into A. Then a characteristic starting in A must remain in A, wherefore there exists at least one limit cycle F in A. By § 5 it will follow that F must loop A, i.e., that its interior cannot lie wholly in A.

By constructing such curves Liénard and Levinson have shown the existence of limit cycles for quite general equations. Details of their methods and further references will be found in the Brown University notes on Non-Linear Mechanics.*

5. Poincaré's Index

Consider a simple closed Jordan curve C in the (x, y) plane passing through no singularity of $V = (P, Q)$. C may be represented in the form

$$x = x(s), \qquad y = y(s); \qquad (s_0 \leq s \leq s_1) \tag{1}$$

$$x(s_0) = x(s_1), \qquad y(s_0) = y(s_1) \tag{2}$$

where $x(s)$ and $y(s)$ are continuous ($s_0 \leq s \leq s_1$), and (2) does not hold for any other pair of values of s. We can stipulate further that as s increases C be described in the positive direction.

It can be shown without difficulty that there exists a unique function $\theta(s)$, continuous for $s_0 \leq s \leq s_1$, and satisfying

$$-\pi < \theta(s_0) \leq \pi$$

$$\sin \theta(s) = \frac{Q(x(s), y(s))}{\sqrt{P^2 + Q^2}}$$

$$\cos \theta(s) = \frac{P(x(s), y(s))}{\sqrt{P^2 + Q^2}} \tag{3}$$

We then define the *index* of the curve C (with respect to V) by

$$I(C) = \frac{\theta(s_1) - \theta(s_0)}{2\pi} \tag{4}$$

Clearly $I(C)$ is an integer.

---

* K. O. Friedricks, P. LeCorbeiller, N. Levinson, and J. J. Stoker, *Non-Linear Mechanics* (mimeographed notes), Brown University, Providence, 1943.

It can further be shown that $I(C)$ is independent of the parametric representation (1) of C. It has the following topological meaning. Let us regard

$$P = P(x, y), \qquad Q = Q(x, y)$$

as a many-one mapping of the x-y plane onto the P-Q plane. Under this mapping C goes into a certain closed curve C' not passing through the origin of the P-Q plane. Then $I(C)$ is the number of times C' circles the origin in the positive direction.

The following properties of the index, though by no means trivial to prove, are fairly obvious intuitively:

(*a*) A simple closed curve C containing in its interior no singularity of V has index 0.

(*b*) A cyclic characteristic F of V has index 1. Hence it follows that a cyclic characteristic must contain at least one singularity.

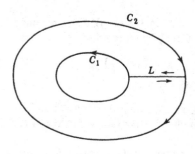

Fig. 5

Suppose now C_1 and C_2 are two concentric curves containing no singularity of V between them. Then they have the same index. For by introducing a suitable cut L joining C_1 and C_2 (Fig. 5), we can define a closed curve Γ containing no singularity whose index is therefore 0. But since the changes in θ along L and back cancel, we have

$$I(\Gamma) = I(C_1) - I(C_2)$$

i.e., C_1 and C_2 have the same index.

In particular, if A is an isolated singularity of V, every simple closed curve about A containing no singularity but A will have the same index. This common index we call the *index of A*. It can be shown without difficulty that if all the singularities of V are isolated, the index of any curve C is equal to the sum of the indices of the singularities it contains. The formula of function theory giving the number N of zeros of an analytic function $f(z)$ inside a curve C as a certain contour integral

$$N = \frac{1}{2\pi i} \int_C \frac{f'(z)}{f(z)} \, dz = \frac{1}{2\pi} \Delta_C \arg (f(z))$$

is a special case of this result.

If F is a cyclic characteristic of V, and all the singularities of V are isolated, then F contains a finite number of singularities the sum of whose indices is 1.

6. Orbital Stability of Limit Cycles

By Theorems 1 and 2, if C is a positive half-characteristic remaining in a closed bounded region $R \subset D$ and C' contains one regular point, then C approaches C' spirally from either the inside or outside. We shall now prove that *every* characteristic starting sufficiently near C' on the same side as C will also approach C' spirally.

Theorem 3. Let C approach its limit set C' spirally from the inside [outside], and let C' contain a regular point A. Then for $\epsilon > 0$ there exists $\delta > 0$ such that every positive half-characteristic starting inside [outside] C', and distant $< \delta$ from C', remains distant $< \epsilon$ from C' and approaches C' spirally.

Proof. Since the singularities of V are isolated, there exists an $\eta > 0$ such that there is no singularity not in C' but distant $< \eta$ from it. Take $\epsilon < \eta$. Let L be a segment without contact through A. Then there exist two points of intersection A_1 and A_2 of C and L, such that the part of C between A_1 and A_2 is everywhere distant $< \epsilon$ from C'. Take δ as the minimum distance of the arc $\widehat{A_1 A_2}$ from C'. Then any characteristic Γ starting at a point distant $< \delta$ from C' and on the same side as C is in the region R bounded by C', $\overline{A_1 A_2}$, and $\widehat{A_1 A_2}$. But Γ cannot get out of R, wherefore $\Gamma' \subset R$. Suppose Γ does not have a limit point on C'. Then Γ' is a limit cycle. If the interior of Γ' lies in R, there must be a singularity in R, a contradiction. Otherwise Γ' must loop the doubly connected region R; which again is impossible, since C would have to cross Γ' to reach C'. Therefore Γ must have at least one limit point on C'. If Γ approached a singularity of C', C could not approach arbitrarily close to C'. Therefore Γ' must contain a regular point B of C'.

Let M be a segment without contact through B; then the intersections of C and Γ on M alternate. Hence by an obvious argument C and Γ have the same limit set, whence the theorem.

Suppose in particular that C' is a cyclic characteristic F. Then F is seen to be stable in a sense analogous to that of Liapunov for singularities. We are thus led to the following definition:

Definition 2. A cyclic characteristic F is said to be orbitally stable if for every $\epsilon > 0$ there exists a $\delta > 0$ such that every positive half-characteristic starting at a point distant $< \delta$ from F is defined for all increasing t and remains distant $< \epsilon$ from F.

(There are various other definitions of stability for characteristics, so we distinguish ours as *orbital* stability.)

Theorem 4. *A necessary and sufficient condition that the cyclic characteristic F be orbitally stable is that for both the interior and exterior of F either:*

(1) *A characteristic approaches F as limit cycle t → + ∞, or*

(2) *There exist cyclic characteristics arbitrarily close to F.*

Proof. From the preceding theorem, sufficiency immediately follows. Suppose F is orbitally stable and on the inside (say) there are no cyclic characteristics or singularities distant $\leq \epsilon$ from F, for some ϵ. Then a positive half-characteristic starting at a point distant $< \delta(\epsilon)$ from F must have F as limit cycle. Hence the theorem.

7. Index of Simple Singularities

We shall finish this chapter by determining the index of any simple singularity and deriving certain applications. We observe first

Lemma 1. *If the simple closed curve C passes through no singularity of either of the vector fields V_1 and V_2, and for no point of C*

$$|\arg V_1 - \arg V_2| = n\pi, \quad n \text{ an integer} > 0 \tag{1}$$

then C has the same index for V_1 and V_2.

For let $\theta_1(P)$, $\theta_2(P)$ (P on C) be the function of § 5 for V_1 and V_2 respectively. Then setting

$$\phi(P) = \theta_1(P) - \theta_2(P)$$

since $|\phi(P)|$ can never take on the value $n\pi$ ($n = 1, 2, \cdots$), it follows that the change of ϕ around C which is a multiple of 2π must be 0

$$\Delta_C\phi = \Delta_C\theta_1 - \Delta_C\theta_2 = 0$$

Hence the lemma.

In particular, let $V(x, y)$ have a simple singularity at $(0, 0)$, and $\bar{V}(x, y)$ be the corresponding linear field. Then for a sufficiently small circle about $(0, 0)$

$$|\arg V - \arg \bar{V}| \leq \frac{\pi}{2}$$

wherefore $(0, 0)$ has the same index for both V and $\bar{V}$. We have therefore only to discuss the index of the singularity at $(0, 0)$ of

$$\begin{aligned} P(x, y) &= ax + by \\ Q(x, y) &= cx + dy \end{aligned} \quad \text{where} \quad \begin{vmatrix} a & b \\ c & d \end{vmatrix} \neq 0 \tag{2}$$

This will be the index of the circle $C: \text{r} = 1$.

Now V cannot have the same direction at two different points (x_1, y_1); (x_2, y_2) on C; for otherwise we would have

$$ax_1 + by_1 = k(ax_2 + by_2)$$
$$cx_1 + dy_1 = k(cx_2 + dy_2), \qquad k > 0 \qquad (3)$$
$$x_1{}^2 + y_1{}^2 = x_2{}^2 + y_2{}^2$$

and the equations (3) have only the solution $k = 1$, $x_1 = x_2$, $y_1 = y_2$. Parametrize C by

$$x = \cos \alpha, \qquad y = \sin \alpha, \qquad 0 \leq \alpha \leq 2\pi$$

Then the angle θ of V is a monotonic function $\theta(\alpha)$. Hence C must have index either $+1$ or -1.

But θ decreases or increases with $\tan \theta(\alpha)$, and

$$f(\alpha) = \tan \theta(\alpha) = \frac{dy}{dx} = \frac{Q}{P} = \frac{c \cos \alpha + d \sin \alpha}{a \cos \alpha + b \sin \alpha}$$

$$f'(\alpha) = \frac{ad - bc}{(a \cos \alpha + b \sin \alpha)^2} \qquad (4)$$

Therefore $f(\alpha)$ and $\theta(\alpha)$ increase or decrease as $\begin{vmatrix} a & b \\ c & d \end{vmatrix}$ is positive. or negative. Hence the theorem:

Theorem 5. A simple singularity has index -1 *if it is a saddle point,* $+1$ *otherwise.*

For the condition that $(0, 0)$ be a saddle point is precisely that

$$\begin{vmatrix} a & b \\ c & d \end{vmatrix} < 0$$

Suppose now that the system

$$\frac{dx}{dt} = P(x, y), \qquad \frac{dy}{dt} = Q(x, y) \qquad (5)$$

has only simple singularities. If these are all saddle points, there can be no cyclic characteristic. A cyclic characteristic must enclose an odd number of singularities, say $2n + 1$, of which n are saddle points and $n + 1$ are centers, nodes, or spiral points. Finally we are enabled to say that in certain regions there can be no cyclic characteristic.

References

R. Bellman, *Stability Theory of Differential Equations*, McGraw-Hill, New York, 1953.

L. Bieberbach, *Theorie der Differentialgleichungen*, Berlin, 1926; reprinted by Dover, New York, 1944.

L. Bieberbach, *Theorie der gewöhnlichen Differentialgleichungen auf funktionentheoretischer Grundlage dargestellt*, Springer, Berlin, 1953.

E. A. Coddington and N. Levinson, *Theory of Ordinary Differential Equations*, McGraw-Hill, New York, 1955.

E. L. Ince, *Ordinary Differential Equations*, London, 1927; reprinted by Dover, New York, 1944.

E. Kamke, *Differentialgleichungen reeller Funktionen*, Leipzig, 1930; reprinted by Chelsea, New York, 1947.

S. Lefschetz, *Differential Equations: Geometric Theory*, Interscience, New York, 1957.

S. Lefschetz, *Lectures on Differential Equations*, Princeton University Press, Princeton, 1946.

V. V. Nemytskii and V. V. Stepanov, *Qualitative Theory of Differential Equations* (Russian), 2nd ed, Gos. Izdat. Tex. Teor. Literatura, Moscow, 1949.

G. Sansone, *Equazioni differenziali nel campo reale*, Part 1, 2nd ed, Nicola Zanichelli Editore, Bologna, 1948; Part 2, 2nd ed, Bologna, 1949.

E. C. Titchmarsh, *Eigenfunction Expansions Associated with Second-Order Differential Equations*, Oxford University Press, Oxford, 1946.

G. Valiron, *Equations fonctionnelles; applications* (Cours d'analyse II), Masson et cie, Paris, 1945.

117

Index

A CATALOG OF SELECTED
DOVER BOOKS
IN SCIENCE AND MATHEMATICS

Astronomy

CHARIOTS FOR APOLLO: The NASA History of Manned Lunar Spacecraft to 1969, Courtney G. Brooks, James M. Grimwood, and Loyd S. Swenson, Jr. This illustrated history by a trio of experts is the definitive reference on the Apollo spacecraft and lunar modules. It traces the vehicles' design, development, and operation in space. More than 100 photographs and illustrations. 576pp. 6 3/4 x 9 1/4.　　　　0-486-46756-2

EXPLORING THE MOON THROUGH BINOCULARS AND SMALL TELESCOPES, Ernest H. Cherrington, Jr. Informative, profusely illustrated guide to locating and identifying craters, rills, seas, mountains, other lunar features. Newly revised and updated with special section of new photos. Over 100 photos and diagrams. 240pp. 8 1/4 x 11.　　　　0-486-24491-1

WHERE NO MAN HAS GONE BEFORE: A History of NASA's Apollo Lunar Expeditions, William David Compton. Introduction by Paul Dickson. This official NASA history traces behind-the-scenes conflicts and cooperation between scientists and engineers. The first half concerns preparations for the Moon landings, and the second half documents the flights that followed Apollo 11. 1989 edition. 432pp. 7 x 10.

0-486-47888-2

APOLLO EXPEDITIONS TO THE MOON: The NASA History, Edited by Edgar M. Cortright. Official NASA publication marks the 40th anniversary of the first lunar landing and features essays by project participants recalling engineering and administrative challenges. Accessible, jargon-free accounts, highlighted by numerous illustrations. 336pp. 8 3/8 x 10 7/8.　　　　0-486-47175-6

ON MARS: Exploration of the Red Planet, 1958-1978--The NASA History, Edward Clinton Ezell and Linda Neuman Ezell. NASA's official history chronicles the start of our explorations of our planetary neighbor. It recounts cooperation among government, industry, and academia, and it features dozens of photos from Viking cameras. 560pp. 6 3/4 x 9 1/4.　　　　0-486-46757-0

ARISTARCHUS OF SAMOS: The Ancient Copernicus, Sir Thomas Heath. Heath's history of astronomy ranges from Homer and Hesiod to Aristarchus and includes quotes from numerous thinkers, compilers, and scholasticists from Thales and Anaximander through Pythagoras, Plato, Aristotle, and Heraclides. 34 figures. 448pp. 5 3/8 x 8 1/2.

0-486-43886-4

AN INTRODUCTION TO CELESTIAL MECHANICS, Forest Ray Moulton. Classic text still unsurpassed in presentation of fundamental principles. Covers rectilinear motion, central forces, problems of two and three bodies, much more. Includes over 200 problems, some with answers. 437pp. 5 3/8 x 8 1/2.　　　　0-486-64687-4

BEYOND THE ATMOSPHERE: Early Years of Space Science, Homer E. Newell. This exciting survey is the work of a top NASA administrator who chronicles technological advances, the relationship of space science to general science, and the space program's social, political, and economic contexts. 528pp. 6 3/4 x 9 1/4.

0-486-47464-X

STAR LORE: Myths, Legends, and Facts, William Tyler Olcott. Captivating retellings of the origins and histories of ancient star groups include Pegasus, Ursa Major, Pleiades, signs of the zodiac, and other constellations. "Classic." – *Sky & Telescope.* 58 illustrations. 544pp. 5 3/8 x 8 1/2.　　　　0-486-43581-4

A COMPLETE MANUAL OF AMATEUR ASTRONOMY: Tools and Techniques for Astronomical Observations, P. Clay Sherrod with Thomas L. Koed. Concise, highly readable book discusses the selection, set-up, and maintenance of a telescope; amateur studies of the sun; lunar topography and occultations; and more. 124 figures. 26 halftones. 37 tables. 335pp. 6 1/2 x 9 1/4.　　　　0-486-42820-6

Browse over 9,000 books at www.doverpublications.com

Chemistry

MOLECULAR COLLISION THEORY, M. S. Child. This high-level monograph offers an analytical treatment of classical scattering by a central force, quantum scattering by a central force, elastic scattering phase shifts, and semi-classical elastic scattering. 1974 edition. 310pp. 5 3/8 x 8 1/2. 0-486-69437-2

HANDBOOK OF COMPUTATIONAL QUANTUM CHEMISTRY, David B. Cook. This comprehensive text provides upper-level undergraduates and graduate students with an accessible introduction to the implementation of quantum ideas in molecular modeling, exploring practical applications alongside theoretical explanations. 1998 edition. 832pp. 5 3/8 x 8 1/2. 0-486-44307-8

RADIOACTIVE SUBSTANCES, Marie Curie. The celebrated scientist's thesis, which directly preceded her 1903 Nobel Prize, discusses establishing atomic character of radioactivity; extraction from pitchblende of polonium and radium; isolation of pure radium chloride; more. 96pp. 5 3/8 x 8 1/2. 0-486-42550-9

CHEMICAL MAGIC, Leonard A. Ford. Classic guide provides intriguing entertainment while elucidating sound scientific principles, with more than 100 unusual stunts: cold fire, dust explosions, a nylon rope trick, a disappearing beaker, much more. 128pp. 5 3/8 x 8 1/2. 0-486-67628-5

ALCHEMY, E. J. Holmyard. Classic study by noted authority covers 2,000 years of alchemical history: religious, mystical overtones; apparatus; signs, symbols, and secret terms; advent of scientific method, much more. Illustrated. 320pp. 5 3/8 x 8 1/2. 0-486-26298-7

CHEMICAL KINETICS AND REACTION DYNAMICS, Paul L. Houston. This text teaches the principles underlying modern chemical kinetics in a clear, direct fashion, using several examples to enhance basic understanding. Solutions to selected problems. 2001 edition. 352pp. 8 3/8 x 11. 0-486-45334-0

PROBLEMS AND SOLUTIONS IN QUANTUM CHEMISTRY AND PHYSICS, Charles S. Johnson and Lee G. Pedersen. Unusually varied problems, with detailed solutions, cover of quantum mechanics, wave mechanics, angular momentum, molecular spectroscopy, scattering theory, more. 280 problems, plus 139 supplementary exercises. 430pp. 6 1/2 x 9 1/4. 0-486-65236-X

ELEMENTS OF CHEMISTRY, Antoine Lavoisier. Monumental classic by the founder of modern chemistry features first explicit statement of law of conservation of matter in chemical change, and more. Facsimile reprint of original (1790) Kerr translation. 539pp. 5 3/8 x 8 1/2. 0-486-64624-6

MAGNETISM AND TRANSITION METAL COMPLEXES, F. E. Mabbs and D. J. Machin. A detailed view of the calculation methods involved in the magnetic properties of transition metal complexes, this volume offers sufficient background for original work in the field. 1973 edition. 240pp. 5 3/8 x 8 1/2. 0-486-46284-6

GENERAL CHEMISTRY, Linus Pauling. Revised third edition of classic first-year text by Nobel laureate. Atomic and molecular structure, quantum mechanics, statistical mechanics, thermodynamics correlated with descriptive chemistry. Problems. 992pp. 5 3/8 x 8 1/2. 0-486-65622-5

ELECTROLYTE SOLUTIONS: Second Revised Edition, R. A. Robinson and R. H. Stokes. Classic text deals primarily with measurement, interpretation of conductance, chemical potential, and diffusion in electrolyte solutions. Detailed theoretical interpretations, plus extensive tables of thermodynamic and transport properties. 1970 edition. 590pp. 5 3/8 x 8 1/2. 0-486-42225-9